LONDON MATHEMATICAL SOCIETY LECTURE NOTE SERIES

Managing Editor: Professor J.W.S. Cassels, Department of Pure Mathematics and Mathematical Statistics, University of Cambridge, 16 Mill Lane, Cambridge CB2 1SB, England

The books in the series listed below are available from booksellers, or, in case of difficulty, from Cambridge University Press.

34 Representation theory of Lie groups, M.F. ATIYAH *et al*
36 Homological group theory, C.T.C. WALL (ed)
39 Affine sets and affine groups, D.G. NORTHCOTT
40 Introduction to H_p spaces, P.J. KOOSIS
46 p-adic analysis: a short course on recent work, N. KOBLITZ
49 Finite geometries and designs, P. CAMERON, J.W.P. HIRSCHFELD & D.R. HUGHES (eds)
50 Commutator calculus and groups of homotopy classes, H.J. BAUES
57 Techniques of geometric topology, R.A. FENN
59 Applicable differential geometry, M. CRAMPIN & F.A.E. PIRANI
66 Several complex variables and complex manifolds II, M.J. FIELD
69 Representation theory, I.M. GELFAND *et al*
74 Symmetric designs: an algebraic approach, E.S. LANDER
76 Spectral theory of linear differential operators and comparison algebras, H.O. CORDES
77 Isolated singular points on complete intersections, E.J.N. LOOIJENGA
79 Probability, statistics and analysis, J.F.C. KINGMAN & G.E.H. REUTER (eds)
80 Introduction to the representation theory of compact and locally compact groups, A. ROBERT
81 Skew fields, P.K. DRAXL
82 Surveys in combinatorics, E.K. LLOYD (ed)
83 Homogeneous structures on Riemannian manifolds, F. TRICERRI & L. VANHECKE
86 Topological topics, I.M. JAMES (ed)
87 Surveys in set theory, A.R.D. MATHIAS (ed)
88 FPF ring theory, C. FAITH & S. PAGE
89 An F-space sampler, N.J. KALTON, N.T. PECK & J.W. ROBERTS
90 Polytopes and symmetry, S.A. ROBERTSON
91 Classgroups of group rings, M.J. TAYLOR
92 Representation of rings over skew fields, A.H. SCHOFIELD
93 Aspects of topology, I.M. JAMES & E.H. KRONHEIMER (eds)
94 Representations of general linear groups, G.D. JAMES
95 Low-dimensional topology 1982, R.A. FENN (ed)
96 Diophantine equations over function fields, R.C. MASON
97 Varieties of constructive mathematics, D.S. BRIDGES & F. RICHMAN
98 Localization in Noetherian rings, A.V. JATEGAONKAR
99 Methods of differential geometry in algebraic topology, M. KAROUBI & C. LERUSTE
100 Stopping time techniques for analysts and probabilists, L. EGGHE
101 Groups and geometry, ROGER C. LYNDON
103 Surveys in combinatorics 1985, I. ANDERSON (ed)
104 Elliptic structures on 3-manifolds, C.B. THOMAS
105 A local spectral theory for closed operators, I. ERDELYI & WANG SHENGWANG
106 Syzygies, E.G. EVANS & P. GRIFFITH
107 Compactification of Siegel moduli schemes, C-L. CHAI
108 Some topics in graph theory, H.P. YAP
109 Diophantine analysis, J. LOXTON & A. VAN DER POORTEN (eds)
110 An introduction to surreal numbers, H. GONSHOR
111 Analytical and geometric aspects of hyperbolic space, D.B.A. EPSTEIN (ed)
113 Lectures on the asymptotic theory of ideals, D. REES
114 Lectures on Bochner-Riesz means, K.M. DAVIS & Y-C. CHANG
115 An introduction to independence for analysts, H.G. DALES & W.H. WOODIN
116 Representations of algebras, P.J. WEBB (ed)
117 Homotopy theory, E. REES & J.D.S. JONES (eds)
118 Skew linear groups, M. SHIRVANI & B. WEHRFRITZ

119 Triangulated categories in the representation theory of finite-dimensional algebras, D. HAPPEL
121 Proceedings of *Groups - St Andrews 1985*, E. ROBERTSON & C. CAMPBELL (eds)
122 Non-classical continuum mechanics, R.J. KNOPS & A.A. LACEY (eds)
124 Lie groupoids and Lie algebroids in differential geometry, K. MACKENZIE
125 Commutator theory for congruence modular varieties, R. FREESE & R. MCKENZIE
126 Van der Corput's method of exponential sums, S.W. GRAHAM & G. KOLESNIK
127 New directions in dynamical systems, T.J. BEDFORD & J.W. SWIFT (eds)
128 Descriptive set theory and the structure of sets of uniqueness, A.S. KECHRIS & A. LOUVEAU
129 The subgroup structure of the finite classical groups, P.B. KLEIDMAN & M.W.LIEBECK
130 Model theory and modules, M. PREST
131 Algebraic, extremal & metric combinatorics, M-M. DEZA, P. FRANKL & I.G. ROSENBERG (eds)
132 Whitehead groups of finite groups, ROBERT OLIVER
133 Linear algebraic monoids, MOHAN S. PUTCHA
134 Number theory and dynamical systems, M. DODSON & J. VICKERS (eds)
135 Operator algebras and applications, 1, D. EVANS & M. TAKESAKI (eds)
136 Operator algebras and applications, 2, D. EVANS & M. TAKESAKI (eds)
137 Analysis at Urbana, I, E. BERKSON, T. PECK, & J. UHL (eds)
138 Analysis at Urbana, II, E. BERKSON, T. PECK, & J. UHL (eds)
139 Advances in homotopy theory, S. SALAMON, B. STEER & W. SUTHERLAND (eds)
140 Geometric aspects of Banach spaces, E.M. PEINADOR and A. RODES (eds)
141 Surveys in combinatorics 1989, J. SIEMONS (ed)
142 The geometry of jet bundles, D.J. SAUNDERS
143 The ergodic theory of discrete groups, PETER J. NICHOLLS
144 Introduction to uniform spaces, I.M. JAMES
145 Homological questions in local algebra, JAN R. STROOKER
146 Cohen-Macaulay modules over Cohen-Macaulay rings, Y. YOSHINO
147 Continuous and discrete modules, S.H. MOHAMED & B.J. MÜLLER
148 Helices and vector bundles, A.N. RUDAKOV et al
149 Solitons, nonlinear evolution equations and inverse scattering, M.J. ABLOWITZ & P.A. CLARKSON
150 Geometry of low-dimensional manifolds 1, S. DONALDSON & C.B. THOMAS (eds)
151 Geometry of low-dimensional manifolds 2, S. DONALDSON & C.B. THOMAS (eds)
152 Oligomorphic permutation groups, P. CAMERON
153 L-functions and arithmetic, J. COATES & M.J. TAYLOR (eds)
154 Number theory and cryptography, J. LOXTON (ed)
155 Classification theories of polarized varieties, TAKAO FUJITA
156 Twistors in mathematics and physics, T.N. BAILEY & R.J. BASTON (eds)
157 Analytic pro-*p* groups, J.D. DIXON, M.P.F. DU SAUTOY, A. MANN & D. SEGAL
158 Geometry of Banach spaces, P.F.X. MÜLLER & W. SCHACHERMAYER (eds)
159 Groups St Andrews 1989 volume 1, C.M. CAMPBELL & E.F. ROBERTSON (eds)
160 Groups St Andrews 1989 volume 2, C.M. CAMPBELL & E.F. ROBERTSON (eds)
161 Lectures on block theory, BURKHARD KÜLSHAMMER
162 Harmonic analysis and representation theory for groups acting on homogeneous trees, A. FIGA-TALAMANCA & C. NEBBIA
163 Topics in varieties of group representations, S.M. VOVSI
164 Quasi-symmetric designs, M.S. SHRIKANDE & S.S. SANE
166 Surveys in combinatorics, 1991, A.D. KEEDWELL (ed)
167 Stochastic analysis, M.T. BARLOW & N.H. BINGHAM (eds)
168 Representations of algebras, H. TACHIKAWA & S. BRENNER (eds)
170 Manifolds with singularities and the Adams-Novikov spectral sequence, B. BOTVINNIK
174 Lectures on mechanics, J.E. MARSDEN
175 Adams memorial symposium on algebraic topology 1, N. RAY & G. WALKER (eds)
176 Adams memorial symposium on algebraic topology 2, N. RAY & G. WALKER (eds)
177 Applications of categories in computer science, M.P. FOURMAN, P.T. JOHNSTONE, & A.M. PITTS (eds)
178 Lower K- and L-theory, A. RANICKI
179 Complex projective geometry, G. ELLINGSRUD, C. PESKINE, G. SACCHIERO & S.A. STROMME (eds)

London Mathematical Society Lecture Note Series. 170

Manifolds with Singularities and the Adams-Novikov Spectral Sequence

Boris I. Botvinnik
Department of Mathematics and Statistics
York University, Ontario

CAMBRIDGE UNIVERSITY PRESS
Cambridge, New York, Melbourne, Madrid, Cape Town, Singapore, São Paulo

Cambridge University Press
The Edinburgh Building, Cambridge CB2 2RU, UK

Published in the United States of America by Cambridge University Press, New York

www.cambridge.org
Information on this title: www.cambridge.org/9780521426084

First published 1992

A catalogue record for this publication is available from the British Library

ISBN-13 978-0-521-42608-4 paperback
ISBN-10 0-521-42608-1 paperback

Transferred to digital printing 2005

Contents

Logical interdependence of the sections

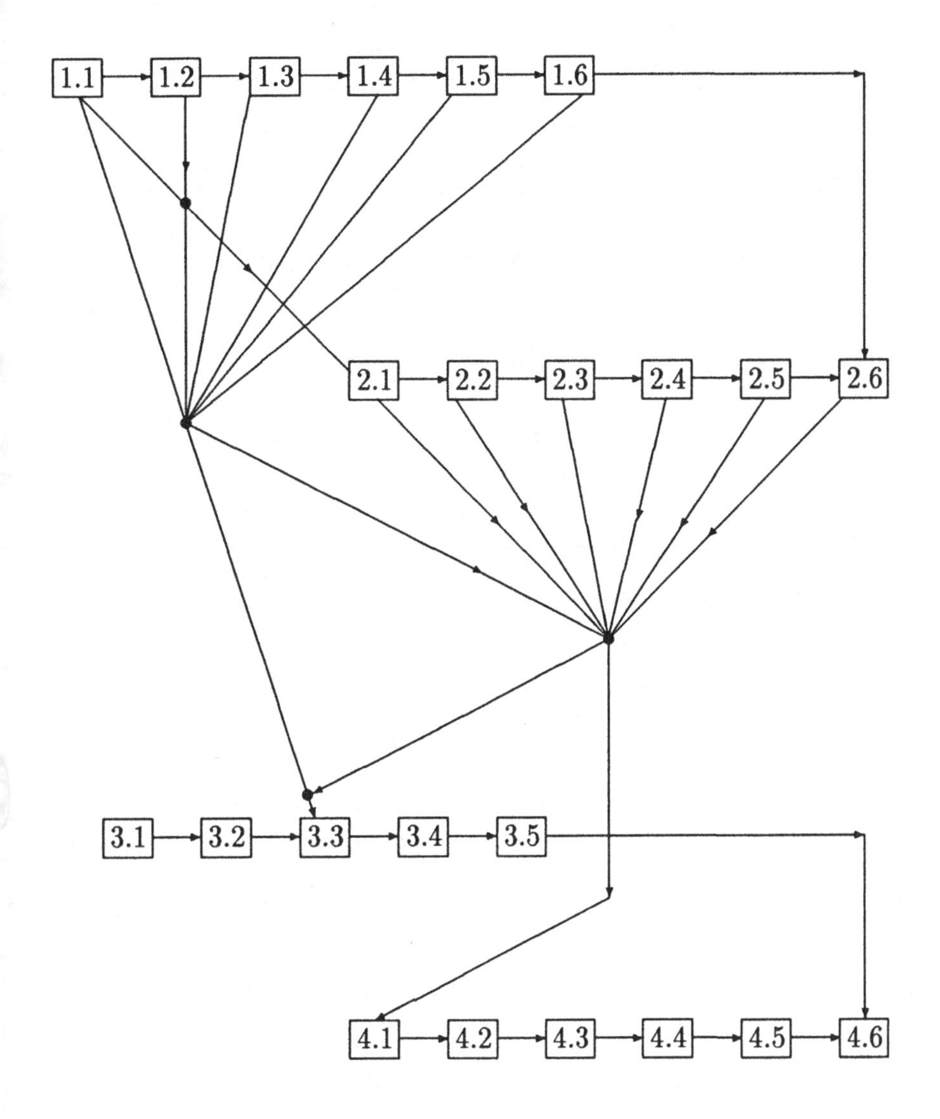

Preface

The purpose of this book is to discuss some natural relations between geometric concepts of Cobordism Theory of manifolds with singularities and the Adams-Novikov spectral sequence.

We begin by motivating this discussion. The central problem of Algebraic Topology has been and continues to be that of obtaining geometrically manageable descriptions of the main algebraic invariants and constructions. For example, let us note the Steenrod problem on realization of integer homology cycles by manifolds. When cycles are presented in terms of manifolds we are able to deal with the cycles by means of Smooth or Piecewise Linear Topology, to apply surgery, to paste and cut and so on. A nice example of successful combining of algebraic and geometric methods may be found in Sullivan's approach to the Hauptvermutung [105]. In particular, Sullivan discovered new geometric objects of Algebraic Topology, namely *manifolds with singularities.* The simplest example is a $(\mathbf{Z}/n)$-manifold (see Figure 0.1, where $n = 4$).

$(\mathbf{Z}/n)$-manifolds present homology cycles with $\mathbf{Z}/n$ coefficients allowing us to apply the machinery of smooth topology. The corresponding (co)-bordism theory naturally represents ordinary (co)-homology theory with coefficients in $\mathbf{Z}/n$.

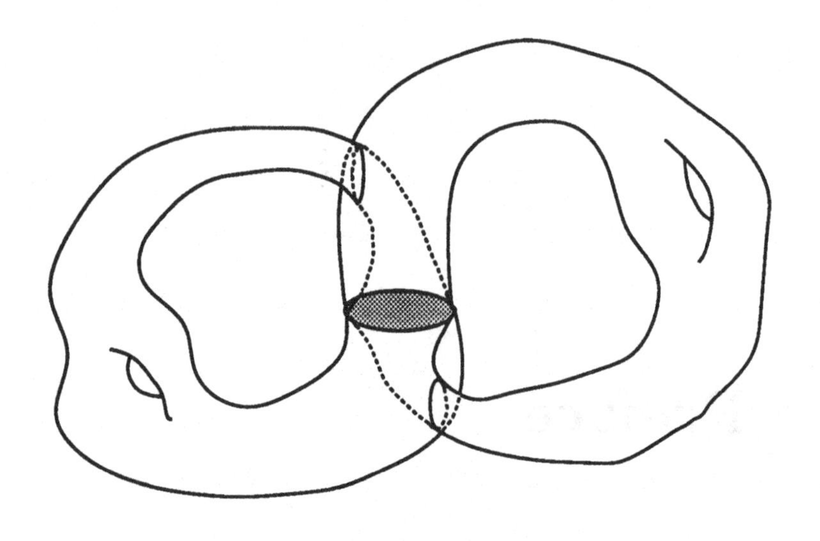

Figure 0.1: $\mathbf{Z}/4$-manifold.

In the early seventies D.Sullivan [106] and N.Baas [11] defined the bordism and cobordism theories of manifolds with singularities in the general case. Then it was made evident that many known homology theories may be realized as bordism theories with singularities. The notion of manifolds with singularities allows us to geometrize the complex and real K-theories, Morava K-theories [93], [115], [116], the Conner-Floyd theory $W(\mathbf{C},2)_*(\cdot)$ [67] and some others whose cycles were not provided with some geometric structure.

Although the notion of *manifold with singularities* is very close to the notion of *ordinary manifold* we have to be much more careful when dealing with the former. For example, the direct product of two manifolds with singularities doesn't possess the structure of such a manifold. Actually there exist some manifolds with singularities which are obstructions to the existence of an *admissible product.* In particular the problem concerning multiplicativity of the ordinary homology theory with $\mathbf{Z}/n$ coefficients may be solved in the same terms. O.K.Mironov [67], [68] was the first to give the general geometric approach to multiplicativity of bordism theories with singularities; see also [9], [16], [70], [116]–[119]. The main constructions pertaining to multiplicativity are described in Chapter 2 of this book.

Now *manifold with singularities* is a common and convenient notion as well as *ordinary manifold.* The corresponding cobordism and bordism theories have been considered from several viewpoints. The connection between the complex cobordism theory with singularities and Formal Group Theory has been found (see [48], [51], [93], [94], [115]–[121]); now a new interesting application has appeared, namely the Theory of Elliptic Genera.

It isn't our purpose to describe all the rich applications of bordism theories with singularities. We would rather concentrate our attention on the connection between geometry of manifolds with singularities and the Adams-Novikov spectral sequence.

Traditionally the Adams-Novikov spectral sequence (ANSS) has been considered as a computational machine allowing us to describe the stable homotopy groups $\pi_* X$ of the given spectrum X in algebraic terms of generators and relations. Actually it can be seen that geometric methods are widely used for description and computation of the Adams-Novikov spectral sequences. For example, the concepts of J-homomorphism and Hopf invariant which originally were geometric tools have been transformed to the powerful chromatic machinery; see [35], [48], [61], [82]–[85]. The chromatic technique allows us to subdivide the Adams-Novikov spectral sequence for a sphere into parts, each of which is determined by the corresponding Morava K-theory $k\langle n\rangle^*(\cdot)$. So the computational problem is reduced to some particular algebraic and homotopy problems.

We note that consideration of the Adams-Novikov spectral sequence for spheres is not the subject of this book. For detailed information refer to D.Ravenel's book [84].

The main topological object which is going to be considered here is the symplectic cobordism ring MSp_*. Actually we are going to examine the Adams-Novikov spectral sequence for this ring. We shall not regard $ANSS$ as a computational tool only, but as a mathematical object provided with rich algebraic and geometric structures. Particular attention will be paid to finding and describing the above geometric structure.

V.Vershinin has gone rather far in the computation of this Adams-Novikov spectral sequence [109]–[113]. Our intention is to use his results and constructions widely.

Geometric methods are also very useful to deal with the Adams-Novikov spectral sequence for the ring MSp_*. For example, Two-valued Formal Group Theory (see V.Buchstaber [26], [27]) describes the ring

$$\text{Л}_* = Hom^*_{\mathbf{A}^{MU}}\left(MU^*(MSp), MU^*\right),$$

which is the zero line of $E_2^{*,*}$, in various terms. We emphasize that application of Two-valued Formal Group Theory is of substantial help to compute this spectral sequence; see [109]–[113].

The first necessary step to start dealing with the Adams-Novikov spectral sequence for a spectrum X is to present a particular *Adams resolution* for this spectrum. Conventionally this is constructed in a standard category of spectra. Sometimes the Adams resolution as well as the Adams-Novikov spectral sequence can be constructed directly from geometric consideration of some notions of cobordism with singularities. To clarify the above statement we consider the following.

Example. The *Adams resolution for the spectrum* MSU can be constructed as follows. Suppose θ_1 is the generator of the group $MSU_1 = \mathbf{Z}/2$, and P is the framed circle presenting the element θ_1 . According to Mironov [67], the bordism theory with θ_1-singularity $MSU^*_{\theta_1}(\cdot)$ is isomorphic to the *Conner-Floyd cohomology theory* $W(\mathbf{C},2)^*(\cdot)$ (see Stong [103]). We note that the Bockstein-Sullivan exact sequence

$$\cdots \longrightarrow MSU^*(\cdot) \xrightarrow{\cdot\theta_1} MSU^*(\cdot) \xrightarrow{\pi} MSU^*_{\theta_1}(\cdot) \xrightarrow{\partial} MSU^*(\cdot) \rightarrow \cdots$$

induces the diagram of the classifying spectra:

$$\begin{array}{ccccccc} MSU & \xleftarrow{\cdot\theta_1} & MSU & \xleftarrow{\cdot\theta_1} & MSU & \xleftarrow{\cdot\theta_1} & \cdots \\ & {\searrow^{\pi}} \quad {\nearrow^{\partial}} & & {\searrow^{\pi}} \quad {\nearrow^{\partial}} & & {\searrow^{\pi}} & \\ & MSU^{\theta_1} & \xrightarrow{\mathfrak{b}} & MSU^{\theta_1} & \xrightarrow{\mathfrak{b}} & \cdots & \end{array} \qquad (0.1)$$

It is obvious that the diagram (0.1) is an Adams resolution of the spectrum MSU in the cohomology theory $MSU^*_{\theta_1}(\cdot)$. In the 2-local

local category this diagram also presents a particular Adams resolution in the Brown-Peterson theory $BP^*(\cdot)$ since the spectrum $MSU^{\theta_1}_{(2)}$ splits into a wedge of the spectra $\Sigma^n BP$.

A complete description of the bordism ring MSU_* in terms of generators and relations was obtained as a result of considering the above Adams-Novikov spectral sequence from geometric and algebraic viewpoints; see [14].

It can be seen now that the Adams-Novikov spectral sequence may be determined by a procedure of resolving singularities. □

It is natural to suppose that such a procedure does exist in the case of several singularities as well. Indeed every given bordism theory $MG_*(\cdot)$ and sequence $\Sigma = (P_1, \cdots, P_k, \cdots)$ of closed manifolds naturally determine the theories $MG_*^{\Sigma\Gamma(k)}(\cdot)$ which are interconnected and are related to the theories $MG_*(\cdot)$, $MG_*^{\Sigma}(\cdot)$ as can be seen in the following diagram:

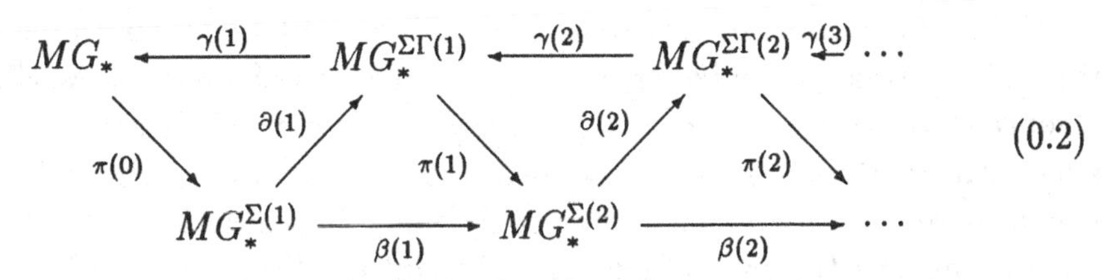

(0.2)

Here the theories $MG_*^{\Sigma(k)}(\cdot)$ split into the sums of the theories $MG_*^{\Sigma}(\cdot)$ and the transformations $\gamma(k)$, $\pi(k)$, $\partial(k)$ will be defined in geometric terms of cutting and gluing the manifolds. For example, the *manifold M in the bordism theory* $MG_*^{\Sigma\Gamma(1)}(\cdot)$ is glued out of the blocks $\gamma_i(M)\times P_i$ in the same way as a *boundary* of a closed manifold with singularities is glued after removing the cones over singularities.

The diagram (0.2), being an exact couple, induces the spectral sequence (it will be called the Σ-*singularities spectral sequence*). This spectral sequence restores the bordism theory $MG_*(\cdot)$ out of the bordism theory with singularities $MG_*^{\Sigma}(\cdot)$.

The top line of the diagram (0.2) represents the filtration of the bordism theory $MG_*(\cdot)$, which may be considered as a geometric analogy of the algebraic filtration generated by powers of the ideal

$$\mathfrak{m} = ([P_1], \cdots, [P_k] \cdots) \subset MG_*.$$

The differentials in the Σ-singularities spectral sequence also have a simple geometric description. The first differential is a direct sum of the Bockstein operators β_k (which are similar to boundary operators on ordinary manifolds). It is very important for our purposes that in several cases the Σ-singularities spectral sequence may be naturally identified with the corresponding Adams-Novikov spectral sequence.

Now let us briefly describe the subject of every chapter. Chapter 1 contains basic geometric constructions of manifolds with singularities. The definition and properties of the Σ-singularities spectral sequence are given at the end of the chapter.

In Chapter 2 we give a geometric construction of a multiplication on bordism theories with singularities. The chapter provides a necessary geometric tool for algebraic considerations.

Chapter 3 is concerned with algebra. First of all we deal with definitions of the Adams-Novikov spectral sequence *(ANSS)* and the Novikov algebraic spectral sequence. Next the symplectic bordism theory with singularities $MSp^{\Sigma}_*(\cdot)$, which has been discovered by V.Vershinin [111], is considered. The coefficient ring MSp^{Σ}_* of this theory is a polynomial ring and the theory $MSp^{\Sigma}_*(\cdot)_{(2)}$ splits into a direct sum of Brown-Peterson theories $BP^*(\cdot)$. So we can identify the corresponding Adams-Novikov spectral sequence with the Σ-singularities spectral sequence. The proof of Vershinin's theorem concerning the theory $MSp^{\Sigma}_*(\cdot)$ is given, since its details will be applied later.

It can be seen that the first differential in the Adams-Novikov spectral sequence splits into a sum of Bockstein operators. So we have a new way to compute the algebra

$$Ext^{*,*}_{\mathbf{A}^{BP}}(BP^*(MSp), BP^*).$$

Some computation is given in Chapter 4. The product structure in the theory $MSp^{\Sigma}_*(\cdot)$ is carefully chosen and the action of Bockstein operators on the generators of the coefficient ring MSp^{Σ}_* is computed.

The result is rather surprising. Indeed, the algebra

$$Ext^{*,*}_{\mathbf{A}^{BP}}(BP^*(MSp^{\theta_1}), BP^*)$$

has a module structure over the symmetric group, so it may be described in terms of representation theory.

So we have tried to come along the way from simple geometric considerations concerning manifolds with singularities up to some computational results describing the structure of the Adams-Novikov spectral sequence.

We use many figures hoping they may be helpful to understand the discussions and arguments.

The list of references doesn't claim completeness.

It is a pleasure to acknowledge Victor M. Buchstaber for helpful comments and valuable critical notes and Vladimir V. Vershinin, Vassily G. Gorbunov, Roin G. Nadiradze for fruitful collaboration and discussions on the subject of this book.

Finally the author would like to thank the staff of the Press Syndicate of the University of Cambridge and the Referee for their unflagging patience and cooperation.

Boris Botvinnik

Chapter 1

Manifolds with singularities

The purpose of this chapter is to describe a procedure restoring an ordinary bordism theory $MG_*(\cdot)$ out of the bordism theory with singularities $MG_*^\Sigma(\cdot)$. This geometric procedure looks like resolving the singularities in the manner of Cusp Theory.

It is our hope that the constructing of the Σ-singularities spectral sequence (Σ-*SSS*) will not get us bogged down in modern Homological Algebra. The main objects for consideration will be manifolds with various geometric structures. The initial point here is the following simple observation. The given bordism theory $MG_*(\cdot)$ and the sequence $\Sigma = (P_1, \cdots, P_n, \cdots)$ of closed manifolds induce not only the bordism theory with singularities $MG_*^\Sigma(\cdot)$, but also the family of intermediate bordism theories $MG_*^{\Sigma\Gamma(k)}(\cdot)$ for $k = 1, 2, \ldots$. A *manifold* M *in the theory* $MG_*^{\Sigma\Gamma(k)}(\cdot)$ (it will be called a $\Sigma\Gamma(k)$-*manifold*) is consistently glued out of the blocks

$$\gamma_\alpha M \times (P_1^{a_1} \times \ldots \times P_n^{a_n} \times \cdots),$$

where $\alpha = (a_1, \ldots, a_n, \ldots)$ is a sequence of nonnegative integers such that $a_1 + \ldots + a_n + \ldots = k$, and $\gamma_\alpha M$ are ordinary manifolds. It is evident that the family of theories $MG_*^{\Sigma\Gamma(k)}(\cdot)$ gives us the filtration of the theory $MG_*(\cdot)$:

$$MG_*(\cdot) \xleftarrow{\gamma(1)} MG_*^{\Sigma\Gamma(1)}(\cdot) \longleftarrow \cdots \longleftarrow MG_*^{\Sigma\Gamma(k)}(\cdot) \xleftarrow{\gamma(k+1)} \cdots$$

This filtration generates the Σ-*SSS*.

Following O.K.Mironov [67], [68] we begin with the definitions of the bordism theories with singularities $MG_*^{\Sigma}(\cdot)$. His constructions seem to have a most clear and obvious form for our purposes. Then we'll define the bordism theories $MG_*^{\Sigma\Gamma(k)}(\cdot)$, $MG_*^{\Sigma(k)}(\cdot)$ from the following diagram in the same manner:

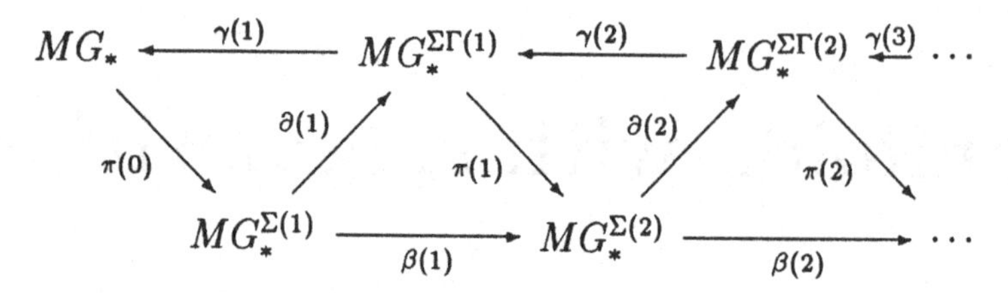

The transformations $\gamma(k)$, $\partial(k)$, $\pi(k)$ will also be defined geometrically, by gluing and cutting the manifolds.

Actually the main part of the chapter (sections 1.2–1.5) contains only some geometric constructions from elementary Cobordism Theory. Almost all the proofs will be given by means of constructing bordisms joining some manifolds. We'll deal with some elementary homological algebra only in section 1.3. We'll define the Σ-*SSS* and prove its simplest properties.

We note also that the Σ-*SSS* was defined by V.V.Vershinin [111] in the one-singularity case, and the general case was described by the present author [15], [16].

1.1 Bordism theories with singularities

Here we define the bordism and cobordism theories with singularities. We wouldn't like to consider an absolutely general case; here we shall be restricted to the following situation.

The starting point is a category of smooth manifolds with a stable G-structure in the stable normal bundle, where G is one of the classic Lie groups. The corresponding bordism and cobordism theories will be denoted by $MG_*(\cdot)$ and $MG^*(\cdot)$. Our main examples will be connected with the cobordism theories $MO^*(\cdot)$, $MSO^*(\cdot)$, $MU^*(\cdot)$, $MSU^*(\cdot)$, $MSp^*(\cdot)$.

Also we suppose that a direct product of the manifolds generates some external product structure in the theories $MG^*(\cdot)$ and $MG_*(\cdot)$ with the usual properties (see [67], and section 2.1). The classifying Thom spectrum for these theories will denoted by MG.

Let us take a sequence $\Sigma = (P_1, \dots, P_n, \dots)$ of closed manifolds. It is supposed below that the sequence Σ is *locally-finite*, i.e. the number sequence $\{\dim P_n\}$ has only infinity as a point of condensation. We denote $\Sigma_k = (P_1, \dots, P_k)$ for every $k = 1, 2, \dots$. It is convenient to denote $P_0 = pt$.

Definition 1.1.1 *We call a manifold M a Σ_k-manifold if there are given the following:*

(i) the partition

$$\partial M = \partial_0 M \cup \partial_1 M \cup \ldots \cup \partial_k M$$

of its boundary ∂M into such a union of manifolds that the intersection

$$\partial_I M = \partial_{i_1} M \cap \ldots \cap \partial_{i_q} M$$

is a manifold for every collection $I = \{i_1, \dots, i_q\} \subset \{0, 1, \dots, k\}$ and its boundary is equal to

$$\partial(\partial_I M) = \bigcup_j (\partial_I M \cap \partial_j M);$$

(ii) the compatible product structures (i.e. diffeomorphisms preserving the stable G-structure)

$$\phi_I : \partial_I M \longrightarrow \beta_I M \times P^I,$$

where $I = \{i_1, \dots, i_q\} \subset \{0, 1, \dots, k\}$,

$$P^I = P_{i_1} \times \ldots \times P_{i_q}.$$

Compatibility here means that if $I \subset J$ and

$$\iota : \beta_I M \longrightarrow \beta_J M$$

is the inclusion, then the map

$$\phi_I \circ \iota \circ \phi_J^{-1} : \beta_I M \times P^J \longrightarrow \beta_I M \times P^I$$

is identical on the direct factor P^I. □

Note 1.1.1 *Now we have defined half-finished manifolds with singularities, to obtain real manifolds with singularities we have to do some identification.* □

Two points x, y of the Σ_k-manifold M are *equivalent* if they belong to the same manifold $\partial_I M$ for some $I \subset \{0, 1, \ldots, k\}$ and

$$pr \circ \phi_I(x) = pr \circ \phi_J(y),$$

where

$$pr : \beta_I M \times P^I \longrightarrow \beta_I M$$

is the projection on the direct factor. The factor-space of the topological space M under this equivalence relation is called *the model of the Σ_k-manifold M* and is denoted by M_Σ.

Indeed it is convenient to deal with Σ_k-manifolds without considering their models. For this we only have to remember consistency of the constructions with the projection

$$\pi : M \longrightarrow M_\Sigma.$$

The *boundary δM of a Σ_k-manifold M* is the manifold $\partial_0 M$. It is also a Σ_k-manifold:

$$\partial_I(\delta M) = \partial_I M \cap \delta M.$$

Manifolds $\beta_I M$ are also Σ_k-manifolds:

$$\partial_j(\beta_I M) = \left\{ \begin{array}{ll} \emptyset & \text{if } j \in I, \\ \beta_{\{j\} \cup I} M \times P_j & \text{otherwise.} \end{array} \right\} \qquad (1.1)$$

Here we denote

$$\beta_I M = \beta_{i_1} \circ \left(\beta_{i_2} \circ \left(\cdots \circ \beta_{i_q} M\right) \cdots\right),$$

if $I = \{i_1, \ldots, i_q\} \subset \{1, \ldots, k\}$.

The pair (M, f) is a *singular Σ_k-manifold of the space pair* (X, Y), if M is a Σ_k-manifold, and

$$f : (M, \delta M) \longrightarrow (X, Y)$$

is such a map that for every index subset $I = \{i_1, \ldots, i_q\} \subset \{1, \ldots, k\}$ the map $f|_{\partial_I M}$ has the following decomposition:

$$f|_{\partial_I M} = f_I \circ pr \circ \phi_I.$$

Here the map

$$pr : \beta_I M \times P^I \longrightarrow \beta_I M$$

is the projection on the direct factor as above and the map

$$f : \beta_I M \longrightarrow X$$

is a continuous one.

Note 1.1.2 *The map f may be also decomposed: $f = f_\Sigma \circ \pi$; here $\pi : M \longrightarrow M_\Sigma$ is the projection, $f_\Sigma : M_\Sigma \longrightarrow X$ is a continuous map. Let us notice that singular Σ_k-manifolds of the point coincide with their topological models.* □

So the bordism theory $MG_*^{\Sigma_k}(\cdot)$ and cobordism theory $MG_\Sigma^*(\cdot)$ of Σ_k-manifolds are well defined. The theories $MG_*^\Sigma(\cdot)$ and $MG_\Sigma^*(\cdot)$ are determined as direct limits of the theories $MG_*^{\Sigma_k}(\cdot)$ and $MG_{\Sigma_k}^*(\cdot)$ respectively.

Theorem 1.1.2 *The theories $MG_*^\Sigma(\cdot)$ and $MG_\Sigma^*(\cdot)$ are extraordinary homology and cohomology theories respectively.*

Proof may be given in a standard manner; it is sufficient to verify that the theories $MG_*^{\Sigma}(\cdot)$ and $MG_{\Sigma}^*(\cdot)$ satisfy the Eilenberg-Steenrod axioms; see [11]. □

Below we will deal mainly with the bordism theories; all the constructions here have a simple geometric interpretation.

Every ordinary manifold may be considered as a Σ_k-manifold with empty set of singularities and the Σ_k-manifold may be considered as a Σ_m-manifold for $m \geq k$. So the following natural transformations are well defined:

$$\pi_k^0 : MG_*(\cdot) \longrightarrow MG_*^{\Sigma_k}(\cdot),$$
$$\pi_n^k : MG_*^{\Sigma_k}(\cdot) \longrightarrow MG_*^{\Sigma_n}(\cdot).$$

The operator $M \longrightarrow \beta_k M$ generates the transformation of degree $-(\dim P_k + 1)$

$$\delta_k : MG_*^{\Sigma_k}(\cdot) \longrightarrow MG_*^{\Sigma_{k-1}}(\cdot).$$

(Note that every manifold $\beta_k M$ is a Σ_{k-1}-manifold by definition.)

The transformations π_k^{k-1}, δ_k connect the theories $MG_*^{\Sigma_k}(\cdot)$ and $MG_*^{\Sigma_{k-1}}(\cdot)$ into the following exact Bockstein-Sullivan triangle:

$$\begin{array}{ccc} MG_*^{\Sigma_{k-1}} & \xleftarrow{\cdot[P_k]} & MG_*^{\Sigma_{k-1}} \\ & {\scriptstyle \pi_k^{k-1}} \searrow \quad \nearrow {\scriptstyle \delta_k} & \\ & MG_*^{\Sigma_k} & \end{array} \tag{1.2}$$

Here we denote the transformation which is generated by direct product (from the right) on the manifold P_k by $\cdot[P_k]$.

1.2 Generalized Bockstein-Sullivan triangle

Now we construct the exact triangle which connects the theories $MG_*(\cdot)$ and $MG_*^{\Sigma}(\cdot)$ for every locally-finite sequence $\Sigma = (P_1, \ldots, P_n, \ldots)$ of closed manifolds. For this we would define a new bordism theory $MG_*^{\Sigma\Gamma(1)}(\cdot)$ which is closely connected with the theories $MG_*^{\Sigma}(\cdot)$ and $MG_*(\cdot)$.

Definition 1.2.1 *The manifold M is called a $\Sigma_k\Gamma(1)$-manifold if there are given*

(i) the partition of the manifolds

$$M = M_1 \cup \cdots \cup M_k, \quad \partial M = \delta M_1 \cup \cdots \cup \delta M_k$$

into a union of manifolds glued along boundaries, i.e. the intersection

$$M_I = M_{i_1} \cap \cdots \cap M_{i_q}$$

is a manifold for every index subset $I = \{i_1, \ldots, i_q\} \subset \{1, \ldots, k\}$ and its boundary is equal to

$$\partial (M_I) = (M_I \cap \partial M) \cup \left(\bigcup_{j \notin I} (M_I \cap M_j) \right);$$

(ii) the compatible product structures

$$\Psi_I : \gamma_I M \longrightarrow \gamma_I M \times P^I,$$

where $\gamma_I M$ are manifolds, $I = \{i_1, \ldots, i_q\} \subset \{1, \ldots, k\}$; compatibility means that the map

$$\Psi_I \circ \iota \circ \Psi_J^{-1} : \gamma_J M \times P^J \longrightarrow \gamma_I M \times P^I$$

is the identity map on the direct factor P^J for every $I \subset J$, where

$$\iota : M_I \longrightarrow M_J$$

is the corresponding inclusion. □

It is evident that a $\Sigma_k\Gamma(1)$-manifold simulates the structure which the part of the boundary of the Σ_k-manifold M,

$$\partial \widetilde{M} = \partial_1 M \cup \ldots \cup \partial_k M$$

has. Its ordinary boundary ∂M is the *boundary* ∂M *of the* $\Sigma\Gamma_k(1)$-*manifold* M; it has this structure by the definition. It should be noted that the manifolds $\gamma_I M$ as well as the manifolds $\beta_I M$ are Σ_k-manifolds.

The map

$$F : (M, \partial M) \longrightarrow (X, Y)$$

is called *the singular* $\Sigma\Gamma_k(1)$-*manifold of the space pair* (X, Y) where M is a $\Sigma\Gamma_k(1)$-manifold, $M = M_1 \cup \ldots \cup M_k$, such that the map $F|_{M_I}$ is decomposed as follows:

$$F|_{M_I} = f_I \circ pr \circ \psi_I$$

for every index subset $I = \{i_1, \ldots, i_q\} \subset \{1, \ldots, k\}$. Here

$$pr : \gamma_I M \times P^I \longrightarrow \gamma_I M$$

is the projection on the direct factor as above, and the map

$$f : \beta_I M \longrightarrow X$$

is a continuous map.

So the bordism theory $MG_*^{\Sigma\Gamma_k(1)}(\cdot)$ is well defined. We define the bordism theory $MG_*^{\Sigma\Gamma(1)}(\cdot)$ as a direct limit of the theories $MG_*^{\Sigma\Gamma_k(1)}(\cdot)$.

Consider the transformation

$$\gamma(1) : MG_*^{\Sigma\Gamma(1)}(\cdot) \longrightarrow MG_*(\cdot),$$

forgetting the $\Sigma\Gamma(1)$-structure and the transformation

$$\partial(1) : MG_*^{\Sigma}(\cdot) \longrightarrow MG_*^{\Sigma\Gamma(1)}(\cdot),$$

defined by the formula:

$$\partial(1) : [(M, f)]_\Sigma \longrightarrow \left[\left(\tilde{\partial} M, f|_{\tilde{\partial} M}\right)\right]_{\Sigma\Gamma(1)},$$

where $\tilde{\partial} M = \partial_1 M \cup \ldots \cup \partial_k M$.

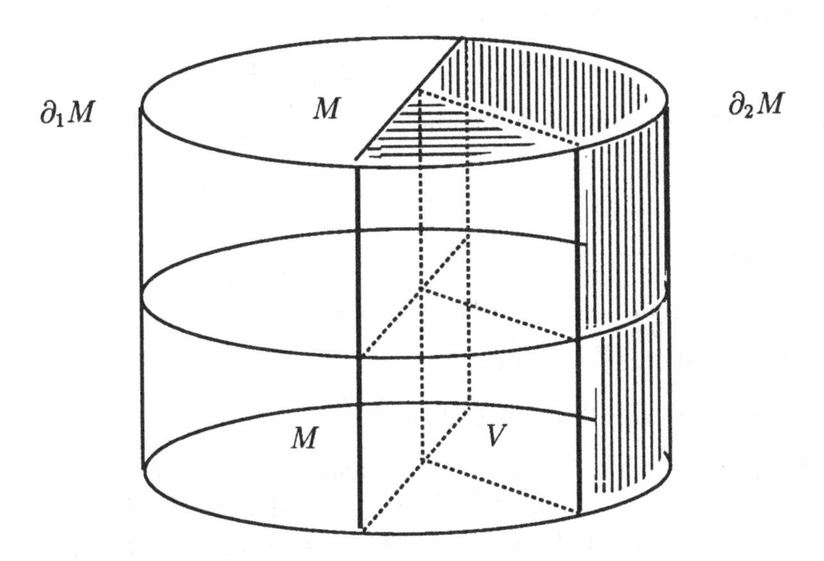

Figure 1.1: $[(L,g)] \in \text{Im}\ \partial(1)$

Theorem 1.2.2 *The following triangle of the theories and transformations is exact:*

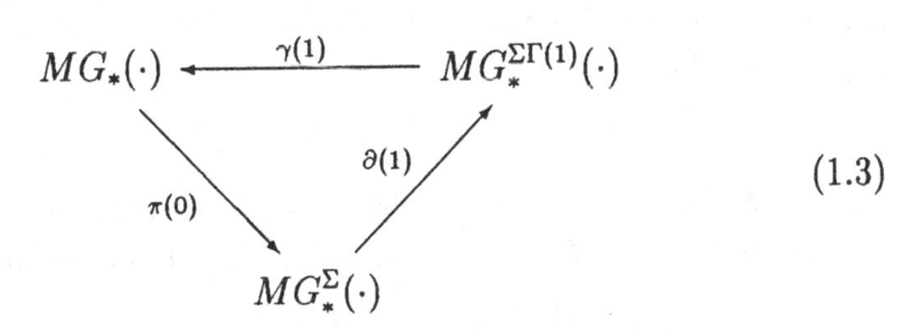

$$\begin{array}{ccc} MG_*(\cdot) & \xleftarrow{\gamma(1)} & MG_*^{\Sigma\Gamma(1)}(\cdot) \\ & {\scriptstyle \pi(0)} \searrow \quad \nearrow {\scriptstyle \partial(1)} & \\ & MG_*^{\Sigma}(\cdot) & \end{array} \tag{1.3}$$

Proof. Let us apply the triangle (1.3) to a space X; we assume here $Y = \emptyset$ for simplicity.

1. Exactness of the vertex $MG_*^{\Sigma}(\cdot)$.

The inclusion $\text{Im}\ \pi \subset \text{Ker}\ \partial(1)$ is obvious. Let (M,f) be a closed singular Σ-manifold, $\partial M = \partial_1 M \cup \cdots \cup \partial_k M$, such that $[(M,f)]_{\Sigma} \in \text{Ker}\ \partial(1)$. Then there exists a singular Σ-manifold (V,G), such that

$$\partial V = \partial M, \quad G|_{\partial M} = f|_{\partial M}.$$

Consider the cylinder (see Figure 1.1)

$$((M \cup -V) \times I, (f \cup -G) \times Id) = (W, H)$$

as a Σ-bordism between the Σ-manifold

$$(M \times \{1\}, f \times \{1\}) = (M, f)$$

and the ordinary manifold

$$((M \cup -V) \times \{0\}, (f \cup -G) \times \{0\}).$$

We obtain that $[(M, f)]_\Sigma \in \text{Im } \pi$.

2. EXACTNESS OF THE VERTEX $MG_*^{\Sigma\Gamma(1)}(\cdot)$.

If (L, g) is a singular $\Sigma\Gamma(1)$-manifold and $[(L, g)] \in \text{Im } \partial(1)$, then there exists a singular Σ-manifold (W, H), such that $\partial W = L$, $H|_{\partial V} = g$; see Figure 1.2. We obtain that $\gamma(1)([(L, g)]) = 0$ by considering (W, H) as an ordinary manifold, i.e. $\text{Im } \partial(1) \subset \text{Ker } \gamma(1)$. The inverse inclusion is obvious.

3. EXACTNESS OF THE VERTEX $MG_*(\cdot)$.

Let (M, f) be an ordinary singular manifold, $[(M, f)] \in \text{Ker } \pi$; then there exists a singular Σ-manifold (V, G) with the boundary $\partial_0 V = M$, $G|_{\partial_0 V} = f$. We have

$$(\partial_1 V \cup \cdots \cup \partial_k V) \cap \partial_0 V = \emptyset$$

because M is a closed manifold. So (V, G) may be considered as a bordism between (M, f) and the following $\Sigma\Gamma(1)$-manifold

$$(\partial_1 V \cup \cdots \cup \partial_k V, G|_{\partial_i V \cup \ldots \cup \partial_k V}).$$

The inclusion $\text{Im } \gamma(1) \subseteq \text{Ker } \pi$ is obvious. □

We have the commutative diagram for every space pair (X, Y)

$$\begin{array}{ccccccc}
\xrightarrow{\delta} MG_*(X,Y) & \xrightarrow{\times[P]} & MG_*(X,Y) & \xrightarrow{\pi} & MG_*^{\Sigma}(X,Y) \xrightarrow{\delta} \\
\downarrow \text{Id} & & \downarrow \omega & & \downarrow \text{Id} \\
\xrightarrow{\delta} MG_*(X,Y) & \xrightarrow{\gamma(1)} & MG_*^{\Sigma\Gamma(1)}(X,Y) & \xrightarrow{\pi(0)} & MG_*^{\Sigma}(X,Y) \xrightarrow{\delta}
\end{array} \tag{1.4}$$

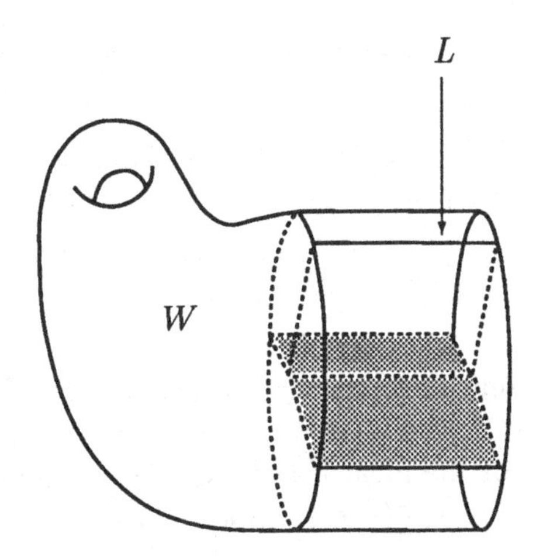

Figure 1.2: Σ-bordism (W, H)

when $\Sigma = (P)$. Here the top line is the Bockstein-Sullivan exact sequence and the transformation ω is generated by the map

$$M \to M \times P.$$

In particular, the theory $MG_*^{\Sigma\Gamma(1)}(\cdot)$ is equivalent to the theory $MG_*(\cdot)$ if $\Sigma = (P)$, where the grading is shifted on $\dim P$.

Corollary 1.2.3 *The bordism theory $MG_*^{\Sigma\Gamma(1)}(\cdot)$ is an extraordinary homology theory.*

Proof. It is sufficient to verify that the theory satisfies the Eilenberg-Steenrod axioms. All these axioms are verified in a standard way. The excision axiom may be verified using the exact triangle (1.3) and the five-lemma. □

Let us consider several examples of the cobordism and bordism theories with singularities.

1. We take a sequence $\Sigma = (P_1, \ldots, P_n, \ldots)$ of closed manifolds in the complex cobordism theory $MU^*(\cdot)$ such that the element $[P_k]$ is a polynomial generator of the ring MU^* in dimension $-2k$, $k = 1, 2, \ldots$.

Then the cobordism theory $MU^*_\Sigma(\cdot)$ coincides with the ordinary cohomology theory $H^*(\cdot;\mathbf{Z})$.

2. We assume $\Sigma=(P)$ for $[P]=\theta_1\in MSU^{-1}$ in the cobordism theory $MSU^*(\cdot)$, where θ_1 is a generator of the group MSU^{-1}. O.K.Mironov [67] noted that the cobordism theory $MSU^*_\Sigma(\cdot)$ coincides with the Conner-Floyd cohomology theory $\mathbf{W}(\mathbf{C},2)^*(\cdot)$. This case was considered in detail in [14].

3. Let us consider a sequence $\Sigma = (P_1,\ldots,P_n,\ldots)$ of closed manifolds in the theory $MU^*(\cdot)$ such that the elements $[P_k]$ give a regular system of generators of the ideal Ker Td $\subset MU^*$ (where Td is Toda genera). Then the theory $MU^*_\Sigma(\cdot)$ is isomorphic to the connected complex K-theory; see [93].

4. Let us consider the oriented cobordism theory $MSO^*(\cdot)$. We put $\Sigma = (P)$, where $[P] = 2$. Then the cobordism theory $MSO^*_\Sigma(\cdot)$ coincides with the w_1-spherical cobordism theory $\mathbf{W}(\mathbf{R},2)^*(\cdot)$ which splits into a wedge of theories $H^*(\cdot;\mathbf{Z}/2)$.

5. Now we consider the integral Brown-Peterson cobordism theory and its generalizations. That is, we choose U-manifolds $P_1,\ldots, P_k,\ldots$, $\dim P_k = 2k$, whose cobordism classes are polynomial generators of the ring MU^*. If $\dim P_k = 2(p^n-1)$ for a prime p, then p divides all the Chern characteristic numbers of the manifolds P_k. Let's take a prime number p and introduce the notations

$$\Sigma^{(0)} = \{P_i \mid i\neq(p^q-1), q=1,2,\ldots\},$$

$$\Sigma^{(n)} = \{P_0, P_{p-1},\ldots,P_{p^n-1}\},$$

where $[P_0]=p, n=1,2,\ldots$. Then the p-localization of the theory

$$MU^*_{\Sigma^{(0)}}(\cdot) = P\langle 0\rangle^*(\cdot)$$

coincides with the Brown-Peterson cohomology theory $BP^*(\cdot)$:

$$P\langle 0\rangle^*(\cdot)\otimes\mathbf{Z}_{(p)} = BP^*(\cdot).$$

The cobordism theory

$$MU^*_{\Sigma^{(n)}}(\cdot) = P\langle n\rangle^*(\cdot)$$

has the following coefficient ring:

$$P\langle n\rangle^* = BP^*/(p, v_1, \dots, v_n),$$

where v_i are the Hazewinkel generators of the ring BP^*; see [8], [9], [51], [67], [68], [117], [118].

1.3 Some complexes of cobordism theories

By definition the Bockstein operator

$$\beta_k : M \longrightarrow \beta_k M$$

determines the transformation

$$\beta_k : MG_*^{\Sigma}(\cdot) \longrightarrow MG_*^{\Sigma}(\cdot)$$

of degree $-(p_k+1)$, where $p_k = \dim P_k$. Definition 1.1. implies the following properties of Bockstein operators:

$$\beta_k \circ \beta_q = 0, \quad \beta_k \circ \beta_q = (-1)^{p_k p_q + 1} \beta_q \circ \beta_k. \tag{1.5}$$

In particular we have the complex

$$MG_{\Sigma}^*(\cdot) \xrightarrow{\beta_k} MG_{\Sigma}^*(\cdot) \xrightarrow{\beta_k} \cdots \xrightarrow{\beta_k} MG_{\Sigma}^*(\cdot) \xrightarrow{\beta_k} MG_{\Sigma}^*(\cdot) \xrightarrow{\beta_k} \cdots \tag{1.6}$$

for every $k = 1, 2, \dots$ and space pair (X, Y). We now introduce the collection of sequences

$$\mathfrak{A} = \{\alpha = (a_1, \dots, a_n, \dots) \mid a_n \geq 0, \textit{all but finitely many } a_n \textit{ are zero}\}.$$

We put

$$L(\alpha)_s(X, Y) = MG^{\Sigma}_{s - \sum_i a_i(p_i+1)}(X, Y)$$

for every $\alpha \in \mathfrak{A}$. So we have that the theory $L(\alpha)_*(\cdot)$is isomorphic to $MG_*^{\Sigma}(\cdot)$ up to degree. The Bockstein operators β_k induce the transformations

$$\beta_k(\alpha) : L(\alpha)_*(\cdot) \longrightarrow L(\alpha_k)_*(\cdot)$$

for every $k = 1, 2, \ldots$. Here $\alpha_k = (a_1, \ldots, a_{k-1}, a_k + 1, a_{k+1}, \ldots)$ for the sequence $\alpha = (a_1, \ldots, a_n, \ldots)$. We define the number

$$\kappa_k(\alpha) = \sum_{i=1}^{k} a_i p_i p_k$$

for every element $\alpha = (a_1, \ldots, a_n, \ldots) \in \mathfrak{A}$. We denote

$$\tilde{\beta}_k(\alpha) = (-1)^{\kappa_k(\alpha)} \beta_k(\alpha).$$

The collection of the graduated groups and transformations

$$\{L(\alpha)_*(X, Y); \beta_1(\alpha), \ldots, \beta_k(\alpha), \ldots\}_{\alpha \in \mathfrak{A}}$$

may be considered as the *lattice of complexes*; we denote it by $\mathcal{L}^\Sigma(X, Y)$.

Lemma 1.3.1 *The differentials $\tilde{\beta}_k(\alpha)$ are anticommutative in the lattice of the complexes $\mathcal{L}^\Sigma(X, Y)$.*

Proof. Consider some square of the lattice $\mathcal{L}^\Sigma(X, Y)$:

$$\begin{array}{ccc} L(\alpha)_*(\cdot) & \xrightarrow{\beta_t(\alpha)} & L(\alpha_t)_*(\cdot) \\ {\scriptstyle \beta_s(\alpha)} \downarrow & & {\scriptstyle \beta_s(\alpha_t)} \downarrow \\ L(\alpha_s)_*(\cdot) & \xrightarrow{\beta_t(\alpha_s)} & L(\alpha_{t,s})_*(\cdot) \end{array} \tag{1.7}$$

where $\alpha_{t,s} = (\alpha_t)_s = (\alpha_s)_t$. Let $s < t$; then we have:

$$\kappa_s(\alpha) + \kappa_t(\alpha_s) + \kappa_t(\alpha) + \kappa_s(\alpha_t)$$

$$= \sum_{i=1}^{s-1} a_i p_i p_s + \sum_{j=1}^{t-1} a_j p_j p_t + \sum_{k=1}^{s-1} a_k p_k p_s + \left(\sum_{l=1}^{t-1} a_l p_l p_t + p_s p_t \right) \equiv p_s p_t \mod 2$$

We obtain from (1.5) that

$$\tilde{\beta}_t(\alpha_s) \circ \tilde{\beta}_s(\alpha) = -\tilde{\beta}_s(\alpha_t) \circ \tilde{\beta}_t(\alpha). \quad \square$$

Let's denote the total complex of the lattice $\mathcal{L}^{\Sigma}(X,Y)$ by $\mathcal{T}^{\Sigma}(X,Y)$:

$$MG_*^{\Sigma(1)}(\cdot) \xrightarrow{\beta(1)} MG_*^{\Sigma(2)}(\cdot) \to \cdots \to MG_*^{\Sigma(k-1)}(\cdot) \xrightarrow{\beta(k-1)} MG_*^{\Sigma(k)}(\cdot) \to \cdots$$

We denote the collection of the partitionings of the nonnegative number k by

$$\mathfrak{A}_k = \left\{ \alpha = (a_1, \ldots, a_n, \ldots) \in \mathfrak{A} \mid \sum_{i=1}^{\infty} a_i = k \right\}.$$

The collections $\mathfrak{A}_k$ will be used below as the index sets for manifolds and their cobordism classes. Let us agree that $M(\alpha) = \emptyset$ and $[M(\alpha)] = 0$ if the sequence α contains a negative element.

The next simple lemma describes the complex $\mathcal{T}^{\Sigma}(X,Y)$ and follows from the above definitions.

Lemma 1.3.2 1) *There is an equivalence of the homology theories*

$$MG_*^{\Sigma(k)}(\cdot) = \bigoplus_{\alpha \in \mathfrak{A}_{k-1}} MG_*^{\Sigma}(\cdot)$$

for every $k = 1, 2, \ldots$. *If the element* x *has the form*

$$x = \bigoplus_{\alpha \in \mathfrak{A}_{k-1}} x(\alpha) \in MG_s^{\Sigma(k)}(X,Y)$$

then $x(\alpha) \in MG_*^{\Sigma}(X,Y)$, *where*

$$\deg x(\alpha) = s - k - \sum_i a_i p_i.$$

2) *The differential* $\beta(k)$ *acts according to the following formula:*

$$\beta(k) \left(\bigoplus_{\alpha \in \mathfrak{A}_{k-1}} x(\alpha) \right) = \bigoplus_{\sigma \in \mathfrak{A}_k} y(\sigma)$$

for every $k = 1, 2, \ldots$; *here we put*

$$y(\sigma) = \sum_{i=1}^{\infty} (-1)^{\kappa_i(\alpha)} \beta_i x(c_1, \ldots, c_{i-1}, c_i - 1, c_{i+1}, \ldots)$$

for every element $\sigma = (c_1, \ldots, c_n, \ldots) \in \mathfrak{A}_k$. □

The local finiteness property of the sequence Σ implies that the groups $MG_n^{\Sigma(k)}(X,Y)$ are finitely generated if the groups $MG_m^{\Sigma}(X,Y)$ are finitely generated for all m.

1.4 The bordism theories $MG_*^{\Sigma\Gamma(k)}(\cdot)$

Here we define the theories $MG_*^{\Sigma\Gamma(k)}(\cdot)$ to extend the triangle (1.3) up to the diagram

$$\begin{array}{ccccccc}
MG_* & \xleftarrow{\gamma(1)} & MG_*^{\Sigma\Gamma(1)} & \xleftarrow{\gamma(2)} & MG_*^{\Sigma\Gamma(2)} & \xleftarrow{\gamma(3)} & \cdots \\
\pi(0)\searrow & \nearrow\partial(1) & \pi(1)\searrow & \nearrow\partial(2) & \pi(2)\searrow & & \\
 & MG_*^{\Sigma(1)} & \xrightarrow{\beta(1)} & MG_*^{\Sigma(2)} & \xrightarrow{\beta(2)} & \cdots &
\end{array}$$

This diagram induces the Σ-*singularities spectral sequence* which restores the bordism theory $MG_*(\cdot)$ from the theory $MG_*^{\Sigma}(\cdot)$.

Definition 1.4.1 *The manifold M is called a $\Sigma\Gamma(k)$-manifold if there are given*

(i) the partitions

$$M = \bigcup_{\alpha\in\mathfrak{A}_k} M(\alpha), \quad \partial M = \bigcup_{\alpha\in\mathfrak{A}_k} \delta M(\alpha),$$

where $M(\alpha)$ are manifolds, $\delta M(\alpha) = M(\alpha)\cap\partial M$, and the equality

$$\partial M(\alpha) = \delta M(\alpha) \cup \left(\bigcup_{\alpha'\in\mathfrak{A}_k,\alpha'\neq\alpha} (M(\alpha)\cap M(\alpha'))\right)$$

holds for each $M(\alpha)$, and a boundary of the manifold

$$M_{\mathfrak{B}} = \bigcap_{\alpha\in\mathfrak{B}} M(\alpha)$$

is equal to

$$\partial M_{\mathfrak{B}} = \bigcup_{\alpha\notin\mathfrak{B}} (M(\alpha)\cap M_{\mathfrak{B}})$$

for every subset $\mathfrak{B} \subseteq \mathfrak{A}_k$ *having more than one element;*

(ii) compatible product structures, i.e. there are defined diffeomorphisms

$$\psi_{\mathfrak{B}} : M_{\mathfrak{B}} \longrightarrow \gamma_{\mathfrak{B}} M \times P^{\mathfrak{B}},$$

preserving the G-structures of the manifolds for every subset $\mathfrak{B} \subset \mathfrak{A}_k$*; here* $\gamma_{\mathfrak{B}} M$ *are ordinary manifolds; we have introduced the notations*

$$P^\alpha = P_1^{a_1} \times \cdots \times P_n^{a_n} \times \cdots$$

for a sequence $\alpha = (a_1, \ldots, a_n, \ldots)$*, and*

$$P^{\mathfrak{B}} = P_1^{b_1} \times \cdots \times P_n^{b_n} \times \ldots,$$

where

$$b_i = \max_{\alpha \in \mathfrak{B}} \{a_i \mid a_i \textit{ the i-th term of the sequence } \alpha\};$$

the compatibility here means that the map

$$\psi_{\mathfrak{C}} \circ \iota \circ \psi_{\mathfrak{B}}^{-1} : \gamma_{\mathfrak{B}}(M) \times P^{\mathfrak{B}} \longrightarrow \gamma_{\mathfrak{C}}(M) \times P^{\mathfrak{C}}$$

is the identity map on the direct factor $P^{\mathfrak{C}}$ *for every subset* $\mathfrak{C} \subset \mathfrak{B}$ *and the corresponding inclusion* $\iota : M_{\mathfrak{B}} \to M_{\mathfrak{C}}$. □

The boundary of a $\Sigma\Gamma(k)$*-manifold* has the $\Sigma\Gamma(k)$-structure by definition. A *singular* $\Sigma\Gamma(k)$*-manifold for the pair* (X, Y) is defined similarly to the case when $k = 1$. So the theory $MG_*^{\Sigma\Gamma(k)}(\cdot)$ is well defined for every $k = 1, 2, \ldots$.

Let's define the transformation

$$\partial(k) : MG_*^{\Sigma(k)}(\cdot) \longrightarrow MG_*^{\Sigma\Gamma(k)}(\cdot)$$

as follows. Let the element x lie in the group $MG_*^{\Sigma(k)}(X, Y)$; then it has the form

$$x = \bigoplus_\alpha [(M_\alpha, g_\alpha)]_\Sigma,$$

where

$$\dim M_\alpha = s - k - \sum_{i=1}^\infty a_i p_i$$

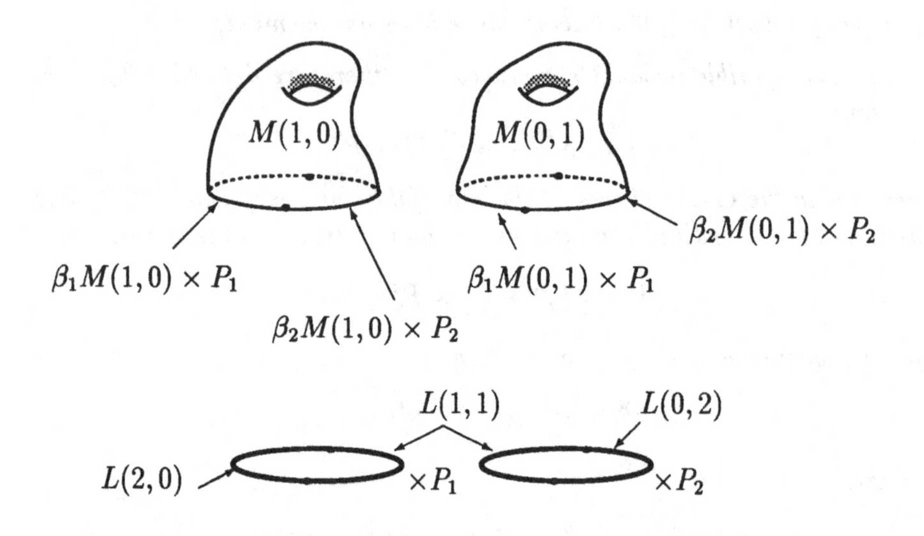

Figure 1.3: $\Sigma\Gamma(k)$-manifold (L, F)

for every $\alpha = (a_1, \ldots, a_n, \ldots)$.

Let us next take the disconnected union of the ordinary singular manifolds

$$(L, F) = \bigcup_{\alpha \in \mathfrak{A}_{k-1}} \left(\tilde{\partial} M_\alpha \times P^\alpha, f_\alpha \circ pr\right),$$

where $\tilde{\partial} M_\alpha = \partial_1 M_\alpha \cup \cdots \cup \partial_m M_\alpha$ as above, and the map

$$pr : \tilde{\partial} M_\alpha \times P^\alpha \to \tilde{\partial} M_\alpha$$

is the projection on the direct factor $f_\alpha = g_\alpha|_{\tilde{\partial} M_\alpha}$.

$$L(\sigma) = \bigcup_{i=1}^{\infty} (-1)^{\kappa_i(\sigma)} \beta_i M(c_1, \ldots, c_{i-1}, c_i - 1, c_{i+1}, \ldots)$$

for every $\sigma = (c_1, \ldots, c_{i-1}, c_i, c_{i+1}, \ldots) \in \mathfrak{A}_k$. So (L, F) has the structure of a $\Sigma\Gamma(k)$-manifold; see Figure 1.3.

It is simple to verify that this construction defines correctly the transformation $\partial(k)$.

The transformation

$$\pi(k) : MG_*^{\Sigma\Gamma(k)}(\cdot) \longrightarrow MG_*^{\Sigma(k+1)}(\cdot)$$

is defined for every $k = 1, 2, \ldots$. That is the Σ-manifold structure is defined on every manifold $\gamma_\alpha M$ for a given $\Sigma\Gamma(k)$-manifold M:

$$M = \bigcup_{\alpha \in \mathfrak{A}_k} M(\alpha), \quad M(\alpha) \overset{\psi_\alpha}{=} \gamma_\alpha M \times P^\alpha.$$

Let us consider the element

$$x = [(M, F)]_{\Sigma\Gamma(k)} \in MG_*^{\Sigma\Gamma(k)}(X, Y),$$

where

$$F|_{M(\alpha)} = f_\alpha \circ pr \circ \psi_\alpha, \quad f_\alpha : \gamma_\alpha M \longrightarrow X.$$

We put $\pi(k)(x) = y$, where

$$y = \bigoplus_{\alpha \in \mathfrak{A}_k} [(\gamma_\alpha M, f_\alpha)]_\Sigma \, .$$

Finally we define the transformation

$$\gamma(k) : MG_*^{\Sigma\Gamma(k)}(\cdot) \longrightarrow MG_*^{\Sigma\Gamma(k-1)}(\cdot)$$

for $k \geq 2$ as follows. We consider the map

$$\xi_k : \mathfrak{A}_k \longrightarrow \mathfrak{A}_{k-1},$$

which is defined on the element $\alpha = (a_1, \ldots, a_n, \ldots) \in \mathfrak{A}_k$ by the formula

$$\xi_k(\alpha) = (a_1, \ldots, a_{t-1}, a_t - 1, a_{t+1}, \ldots) \in \mathfrak{A}_{k-1},$$

where $t = \min\{j \mid a_j \geq 1\}$. Let the map

$$F : (M, \partial M) \longrightarrow (X, Y)$$

be a representative of the element $x \in MG_*^{\Sigma\Gamma(k)}(X, Y)$, where $x = [M]$,

$$M = \bigcup_{\alpha \in \mathfrak{A}_k} M(\alpha), \quad M(\alpha) \overset{\psi_\alpha}{=} \gamma_\alpha M \times P^\alpha.$$

We note that

$$\gamma_\alpha M \times P^\alpha = \gamma_\alpha M \times P_t \times P^\sigma,$$

where $t = t(\alpha)$, $\sigma = \xi_k(\alpha)$. We define the $\Sigma\Gamma(k-1)$-structure on the manifold $L = M$ as follows. We put

$$L(\sigma) = \bigcup_{\alpha \in \xi^{-1}(\sigma)} \gamma_\alpha M$$

for every $\sigma \in \mathfrak{A}_{k-1}$ and then let

$$\gamma(k)(x) = [(L, F)]_{\Sigma\Gamma(k-1)}.$$

Note 1.4.1 *The transformation $\gamma(k)$ depends on the choice of the map ξ_k , which is not symmetric with respect to the manifolds P_k. For example, we consider below the case when $\Sigma = (P_1, P_2)$.* □

Suppose M is a $\Sigma\Gamma(2)$-manifold:

$$M = M\,(2,0) \cup M\,(1,1) \cup M\,(0,2). \tag{1.8}$$

By applying the transformation $\gamma(k)$ we obtain the $\Sigma\Gamma(1)$-manifold $L = L\,(1,0) \cup L\,(0,1) = M$:

$$L\,(1,0) = M\,(2,0)\,, \qquad L\,(0,1) = M\,(1,1) \cup M\,(0,2)\,.$$

We choose another $\Sigma\Gamma(k)$-structure on the manifold L as follows:

$$\widetilde{L}\,(1,0) = M\,(0,2)\,, \qquad \widetilde{L}\,(0,1) = M\,(2,0) \cup M\,(1,1)\,.$$

The bordism joining these two $\Sigma\Gamma(1)$-manifolds is displayed in Figure 1.4.

It also can be shown that in the general case the $\Sigma\Gamma(k-1)$-bordism class of the manifold $\gamma(k)\,(M)$ doesn't depend on the choice of the $\Sigma\Gamma(k-1)$-structure (it need only be compatible with the original $\Sigma\Gamma(k)$-structure on the manifold M).

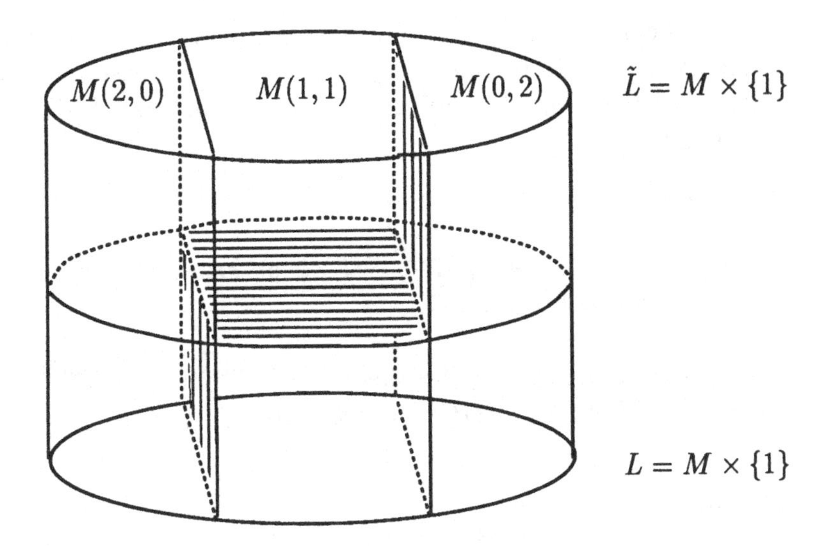

Figure 1.4: Bordism between L and $\tilde{L}$

Theorem 1.4.2 *The following triangle of the bordism theories and transformations is exact:*

$$\begin{array}{ccc} MG_*^{\Sigma\Gamma(k-1)}(\cdot) & \xleftarrow{\gamma(k)} & MG_*^{\Sigma\Gamma(k)}(\cdot) \\ & {\scriptstyle \pi(k-1)} \searrow \quad \nearrow {\scriptstyle \partial(k)} & \\ & MG_*^{\Sigma(k)}(\cdot) & \end{array} \tag{1.9}$$

for every $k = 2, 3, \ldots$. The identity

$$\pi(k) \circ \partial(k) = \beta(k)$$

also holds.

Proof of this theorem will be given in section 1.5. □

Note 1.4.2 *Consider the case when $\Sigma = (P)$. Then the theory $MG_*^{\Sigma\Gamma(k)}(\cdot)$ is naturally isomorphic to the theory $MG_*(\cdot)$ for every $k = 1, 2, \ldots$ by changing the graduation on pk, where $p = \dim P$, and the triangle (1.9) is isomorphic to the Bockstein-Sullivan triangle (1.2).* □

It is not difficult to verify that the theory $MG_*^{\Sigma\Gamma(k)}(\cdot)$ satisfies the Eilenberg-Steenrod axioms for every $k \geq 1$ as in the case of $k = 1$. The excision axiom may be verified by the help of triangle (1.9) and the five-lemma.

So we obtain

Corollary 1.4.3 *The theory $MG_*^{\Sigma\Gamma(k)}(\cdot)$ is an extraordinary homology theory for every $k = 1, 2, \ldots$.* □

The spectra representing the theories $MG_*^{\Sigma\Gamma(k)}(\cdot)$, $MG_*^{\Sigma(k)}(\cdot)$ will be denoted by $MG^{\Sigma\Gamma(k)}$, $MG^{\Sigma(k)}$ respectively.

1.5 Proof of Theorem 1.4.2

The proof will be given by direct constructing of the corresponding bordism manifolds. For simplicity we consider the case when the space Y is empty. When the space Y isn't empty the proof is quite similar. Let's apply the triangle (1.9) to the space X.

1 EXACTNESS OF THE VERTEX $MG_*^{\Sigma(k)}$.

i) Ker $\partial(k) \subseteq$ Im $\pi(k)$. Let $x \in$ Ker $\partial(k)$; it has the form

$$x = \bigoplus_{\alpha \in \mathfrak{A}_{k-1}} [(M(\alpha), F(\alpha))]_\Sigma ,$$

where $M(\alpha)$ are Σ-manifolds and the maps $F(\alpha)|_{\partial_i M(\alpha)}$ are decomposed as follows:

$$F(\alpha)|_{\partial_i M(\alpha)} : \partial_i M(\alpha) \xrightarrow{\phi_i(\alpha)} \beta_i M(\alpha) \times P^\alpha \xrightarrow{pr} \beta_i M(\alpha) \xrightarrow{f_i(\alpha)} X .$$

The definition of the transformation $\partial(k)$ implies that there exists a singular $\Sigma\Gamma(k)$-manifold (V, G) with boundary $(\partial V, G|_{\partial V})$, such that

$$V = \bigcup_{\sigma \in \mathfrak{A}_k} V(\sigma), \quad V(\sigma) \stackrel{\psi_\sigma}{\cong} \gamma_\sigma V \times P^\sigma, \quad \partial V = \bigcup_{\sigma \in \mathfrak{A}_k} \delta V(\sigma).$$

Here we introduce the following notation:

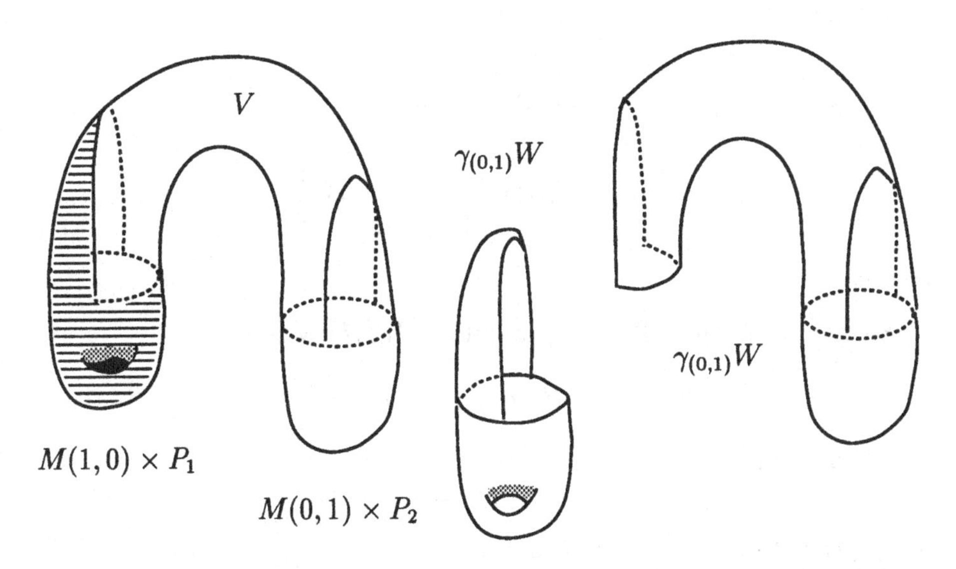

Figure 1.5: $\Sigma\Gamma(k-1)$-manifold W.

$$\delta V(\sigma) = \bigcup_{i=1}^{\infty} (-1)^{\kappa_i(\sigma)} \beta_i M(c_1, \ldots, c_{i-1}, c_i - 1, c_{i+1}, \ldots) \times P^{\sigma},$$

$$G|_{\delta V(\sigma)} = \bigcup_{i=1}^{\infty} (-1)^{\kappa_i(\sigma)} f_i(c_1, \ldots, c_{i-1}, c_i - 1, c_{i+1}, \ldots) \circ pr$$

for every element $\sigma = (c_1, \ldots, c_n, \ldots) \in \mathfrak{A}_k$. Let's consider the manifold

$$V_1 = - \bigcup_{\alpha \in \mathfrak{A}_{k-1}} M(\alpha) \times P^{\alpha}.$$

We glue it with the manifold V along the boundary $\partial V = \partial V_1$ using the equalities

$$\partial_i M(\alpha) \times P^{\alpha} = \beta_i M(\alpha) \times P_i \times P^{\alpha} = (-1)^{\kappa_i(\alpha)} \beta_i M(\alpha) \times P^{\alpha_i},$$

where $\alpha_i = (a_1, \ldots, a_{i-1}, a_i - 1, a_{i+1}, \ldots)$, if $\alpha = (a_1, \ldots, a_i, \ldots)$.

We define the $\Sigma\Gamma(k-1)$-structure on the manifold $W = V \cup_\partial -V_1$ by putting

$$W(\alpha) = \gamma_\alpha W \times P^\alpha$$

for every $\alpha = (a_1, \dots, a_n, \dots) \in \mathfrak{A}_{k-1}$, where

$$\gamma_\alpha W = -M(\alpha) \cup \left(\bigcup_{\sigma \in \xi^{-1}(\alpha)} \gamma_\sigma V \right);$$

see Figure 1.5.

So we obtain the $\Sigma\Gamma(k-1)$-manifold (W, H), where

$$H|_V = G, \quad H|_{M(\alpha)\times P^\alpha} = F(\alpha) \circ pr,$$

where the map

$$pr : M(\alpha) \times P^\alpha \longrightarrow M(\alpha)$$

is the projection on the direct factor.

Also we have that

$$H(\alpha) = H|_{W(\alpha)} = h_\alpha \circ pr \circ \phi_\alpha,$$

where

$$pr : \gamma_\alpha W \times P^\alpha \longrightarrow \gamma_\alpha W, \quad h_\alpha : \gamma_\alpha W \longrightarrow X.$$

We have obtained

$$\pi(k-1)\left([(W,H)]_{\Sigma\Gamma(k-1)}\right) = \bigoplus_{\alpha \in \mathfrak{A}_{k-1}} [(\gamma_\alpha W, h_\alpha)]_\Sigma,$$

by the definition of $\pi(k-1)$.

It is clear that the Σ-manifolds $(\gamma_\alpha W, h_\alpha)$, $(M(\alpha), F(\alpha))$ are bordant for every $\alpha \in \mathfrak{A}_{k-1}$ according to the construction of the manifolds $\gamma_\alpha W$.

ii) <u>$\mathrm{Im}\ \pi(k-1) \subseteq \mathrm{Ker}\ \partial(k)$.</u>

Let $x \in \mathrm{Ker}\ \pi(k-1)$. Let us find the element y, such that $\pi(k-1)(y) = x$; here $y = [(L, H)]$:

$$L = \bigcup_{\alpha \in \mathfrak{A}_{k-1}} L(\alpha), \quad L(\alpha) \overset{\psi_\alpha}{\cong} \gamma_\alpha L \times P^\alpha, \quad H|_{L(\alpha)} = h_\alpha \circ pr \circ \phi_\alpha,$$

where

$$pr : \gamma_\alpha L \times P^\alpha \longrightarrow \gamma_\alpha L, \quad h_\alpha : \gamma_\alpha L \longrightarrow X.$$

If the element x has the form

$$x = \bigoplus_{\alpha \in \mathfrak{A}_{k-1}} [(M(\alpha), F(\alpha))]_\Sigma ,$$

then the next identity is true by the definition of the transformation $\pi(k-1)$:

$$[(M(\alpha), F(\alpha))]_\Sigma = [(\gamma_\alpha L, h_\alpha)]_\Sigma , \quad \alpha \in \mathfrak{A}_{k-1}.$$

Now we take some Σ-bordism $(V(\alpha), G(\alpha))$ between the Σ-manifolds $(M(\alpha), F(\alpha))$ and $(\gamma_\alpha L, h_\alpha)$ for every $\alpha \in \mathfrak{A}_{k-1}$.

Let us glue the cylinder $-L \times I$ with the manifold

$$W = - \bigcup_{\alpha \in \mathfrak{A}_{k-1}} V(\alpha) \times P^\alpha$$

along the boundary by identifying the manifolds (see Figure 1.6)

$$-L(\alpha) \times \{0\} \quad \text{and} \quad \gamma_\alpha L \times P^\alpha \cong V(\alpha) \times P^\alpha.$$

The boundary of the obtained manifold U is decomposed into the union of the next three manifolds:

$$\partial U = -L \times \{1\} \cup \left(\bigcup_{\alpha \in \mathfrak{A}_{k-1}} \left(\bigcup_{i=1}^{\infty} \beta_i V(\alpha) \times P^\alpha \times P_i \right) \right) \cup \bigcup_{\alpha \in \mathfrak{A}_{k-1}} M(\alpha) \times P^\alpha.$$

We denote the second part of the boundary ∂W by S and consider it as a singular $\Sigma\Gamma(k)$-manifold (S, E) by putting

$$\gamma_\sigma S = \bigcup_{i=1}^{\infty} (-1)^{\kappa_i(\sigma)} \beta_i V(c_1, \ldots, c_{i-1}, c_i - 1, c_{i+1}, \ldots),$$

for every $\sigma = (c_1, \ldots, c_n, \ldots) \in \mathfrak{A}_k$, where the map E is defined by the maps $G(\alpha)$,

$$G(\alpha)|_{\partial_i V(\alpha)} : \partial_i V(\alpha) \xrightarrow{\phi_i(\alpha)} \beta_i V(\alpha) \times P^\alpha \xrightarrow{pr} \beta_i V(\alpha) \xrightarrow{g_i(\alpha)} X,$$

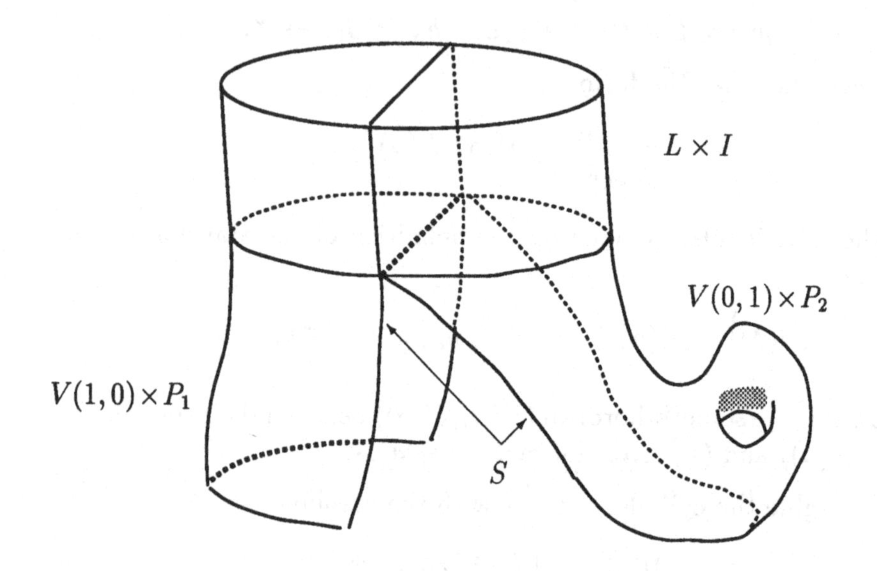

Figure 1.6: $\Sigma\Gamma(k)$-manifold U.

as follows:

$$E|_{\partial_i V(\alpha)\times P^\alpha} : \partial_i V(\alpha) \times P^\alpha \xrightarrow{pr} \partial_i V(\alpha) \xrightarrow{\phi_i(\alpha)\circ pr} \beta_i V(\alpha) \xrightarrow{g_i(\alpha)} X.$$

Let's note the next identity:

$$E|_{S(\sigma)} = \bigcup_{i=1}^{\infty} \left((-1)^{\kappa_i(\sigma)} f_i(c_1,\ldots,c_{i-1},c_i-1,c_{i+1},\ldots) \circ \phi_i(\alpha) \circ pr\right).$$

We obtain $\partial(k)(x) = 0$ by the definition of the transformation $\partial(k)$.

2 Exactness of the vertex $MG_*^{\Sigma\Gamma(k-1)}$.

i) Ker $\pi(k-1) \subseteq$ Im $\gamma(k)$.

Suppose the element $[(L,G)]$ lies in Ker $\pi(k-1)$, where

$$L = \bigcup_{\alpha\in\mathfrak{A}_{k-1}} L(\alpha), \quad L(\alpha) \stackrel{\psi_\alpha}{\cong} \gamma_\alpha L \times P^\alpha,$$

$$G|_{L(\alpha)} : L(\alpha) \xrightarrow{\psi_\alpha} \gamma_\alpha L \times P^\alpha \xrightarrow{pr} \gamma_\alpha L \xrightarrow{g_\alpha} X,$$

and $[(\gamma_\alpha L, g_\alpha)]_\Sigma = 0$ for all $\alpha \in \mathfrak{A}_{k-1}$. We take a singular Σ-manifold $(V(\alpha), H(\alpha))$ for every $\alpha \in \mathfrak{A}_{k-1}$ such that

$$\delta V(\alpha) = \gamma_\alpha L, H|_{\delta V(\alpha)} = g(\alpha).$$

Then we glue the manifold $-V(\alpha) \times P^\alpha$ to the top side of the cylinder $L \times I$ by identifying the manifolds $\gamma_\alpha L \times P^\alpha$ and $-\delta V(\alpha) \times P^\alpha$. We consider the obtained manifold W as the $\Sigma\Gamma(k-1)$-bordism between the $\Sigma\Gamma(k-1)$-manifolds $L \times \{0\}$ and M, where

$$M = \bigcup_{\alpha \in \mathfrak{A}_{k-1}} (\beta_1 V(\alpha) \times P_1 \cup \cdots \cup \beta_m V(\alpha) \times P_i) \times P^\alpha.$$

In general the $\Sigma\Gamma(k-1)$-manifold M is not the image of some $\Sigma\Gamma(k)$-manifold under the action of the transformation $\gamma(k)$. Let us consider the cylinder $M \times I$ and define the $\Sigma\Gamma(k)$-structure on it as follows. Let's introduce the notation

$$\alpha = (a_1, \ldots, a_n, \ldots) \in \mathfrak{A}_k, \quad \alpha_i = (a_1, \ldots, a_{i-1}, a_i - 1, a_{i+1}, \ldots) \in \mathfrak{A}_{k-1}.$$

We put

$$(M \times I)(\tilde{\alpha}) = \bigcup_{\substack{\alpha \in \mathfrak{A}_{k-1} \\ \alpha_i \in \xi^{-1}(\tilde{\alpha})}} (\beta_i V(\alpha) \times P_i) \times P^\alpha \times I$$

for every $\tilde{\alpha} \in \mathfrak{A}_{k-1}$. We glue the cylinder $M \times I$ along the bottom side with the manifold W by identifying the manifolds $M \times \{0\}$ and $M \subseteq W$. As a result we obtain the next $\Sigma\Gamma(k-1)$-manifold:

$$U = \bigcup_{\alpha \in \mathfrak{A}_{k-1}} U(\alpha),$$

where $U(\alpha) = W(\alpha) \cup (M \times I)(\alpha)$; see Figure 1.7.

The correctness of the defined $\Sigma\Gamma(k-1)$-structure on U may be verified directly. The last step is to note that the $\Sigma\Gamma(k)$-structure is well defined on the manifold $M \times I \subset \partial U$. That is, we put

$$M(\sigma) = \beta_t V(\xi_k(\sigma)) \times P^\sigma$$

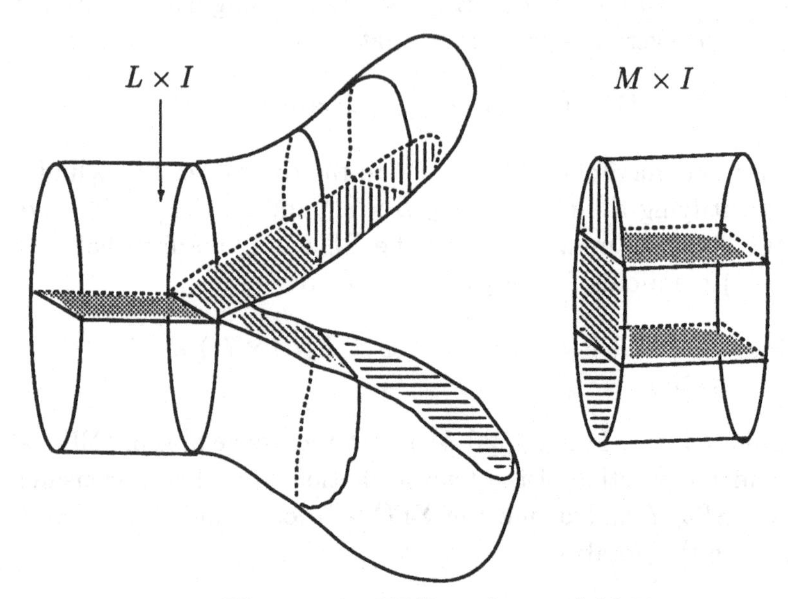

Figure 1.7: $\Sigma\Gamma(k-1)$-manifold U.

for every $\sigma \in \mathfrak{A}_k$, where

$$t = \min\{\, j \mid c_j \geq 1 \,\}, \quad \sigma = (c_1, \ldots, c_n, \ldots).$$

The above construction defines the singular $\Sigma\Gamma(k-1)$-manifold (U, F) having as first boundary (L, G) and as second boundary the singular $\Sigma\Gamma(k-1)$-manifold which lies in the image of the transformation $\gamma(k)$.

ii) Im $\gamma(k) \subseteq$ Ker $\pi(k)$.

Let $x = [(M, F)] \in \operatorname{Im} \gamma(k)$, where

$$M = \bigcup_{\alpha \in \mathfrak{A}_{k-1}} M(\alpha), \quad M(\alpha) \overset{\psi_\alpha}{\cong} \gamma_\alpha M \times P^\alpha,$$

$$F|_{M(\alpha)} : M(\alpha) \xrightarrow{\psi_\alpha} \gamma_\alpha M \times P^\alpha \xrightarrow{pr} \gamma_\alpha M \xrightarrow{f_\alpha} X.$$

Then there exists a $\Sigma\Gamma(k-1)$-bordism (W, H), such that

$$W = \bigcup_{\alpha \in \mathfrak{A}_{k-1}} W(\alpha), \quad W(\alpha) \overset{\phi_\alpha}{\cong} \gamma_\alpha W \times P^\alpha,$$

and also $\partial W = M \cup L$, $H|_M = F$,

$$F|_{W(\alpha)} : W(\alpha) \xrightarrow{\phi_\alpha} \gamma_\alpha W \times P^\alpha \xrightarrow{pr} \gamma_\alpha W \xrightarrow{h_\alpha} X,$$

where $(L, H|_L)$ is a singular $\Sigma\Gamma(k)$-manifold:

$$L = \bigcup_{\sigma \in \mathfrak{A}_k} L(\sigma), \quad L(\sigma) \stackrel{\tau_\sigma}{=} \gamma_\sigma L \times P^\sigma .$$

We consider now the manifold $(L, H|_L)$ as a singular $\Sigma\Gamma(k-1)$-manifold as follows according to the definition of the transformation $\gamma(k)$. We put

$$L(\alpha) = \bigcup_{\sigma \in \xi^{-1}(\alpha)} L(\sigma), \quad \gamma_\alpha L = \bigcup_{\sigma \in \xi^{-1}(\alpha)} (\gamma_\alpha L \times P_t \times P^\alpha),$$

for every $\alpha = (a_1, \ldots, a_n, \ldots) \in \mathfrak{A}_{k-1}$, where

$$t = t_\alpha = \min\{\, j \mid a_j \geq 1\} .$$

In particular the manifold $(\gamma_\alpha W, h_\alpha)$ is a singular Σ-manifold with the boundary

$$\delta(\gamma_\alpha W, h_\alpha) = (\gamma_\alpha M, f_\alpha) .$$

This means that $x \in \operatorname{Ker} \pi(k-1)$.

3 EXACTNESS OF THE VERTEX $MG_*^{\Sigma(k)}$.

i) <u>$\operatorname{Im} \partial(k) \subseteq \operatorname{Ker} \gamma(k)$.</u>

Let $x \in [(M, F)] \in \operatorname{Im} \partial(k)$, where

$$M = \bigcup_{\alpha \in \mathfrak{A}_k} M(\alpha), \quad M(\alpha) \stackrel{\psi_\alpha}{=} \gamma_\alpha M \times P^\alpha,$$

$$F|_{M(\alpha)} : M(\alpha) \xrightarrow{\psi_\alpha} \gamma_\alpha M \times P^\alpha \xrightarrow{pr} \gamma_\alpha M \xrightarrow{f_\alpha} X.$$

There exists a $\Sigma\Gamma(k)$-bordism (V, G) such that $\partial V = M \cup L$, $G|_M = F$ according to the definition of the transformation $\partial(k)$, where the $\Sigma\Gamma(k)$-manifold L is defined by the singular $\Sigma(k)$-manifold

$$\{(S(\sigma), H(\sigma))\}_{\sigma \in \mathfrak{A}_{k-1}}$$

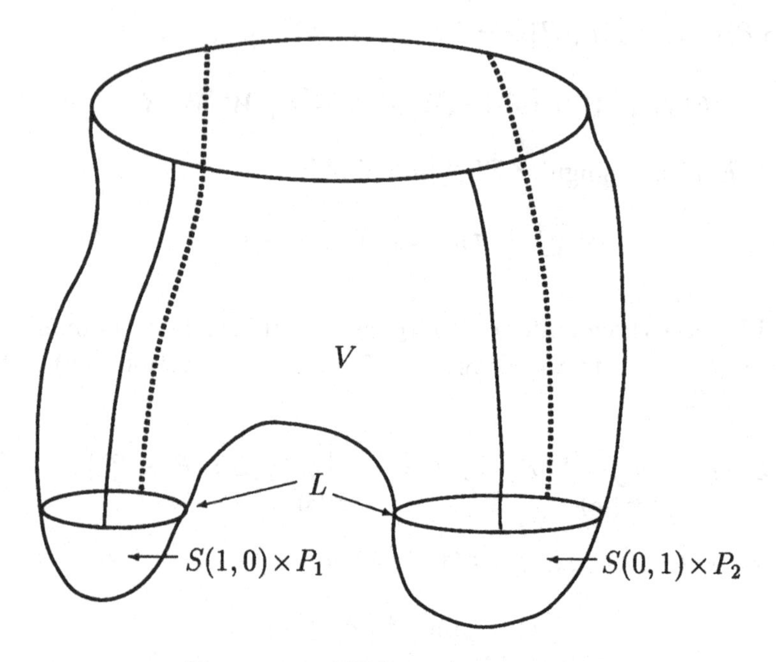

Figure 1.8: $\Sigma\Gamma(k-1)$-manifold W.

as follows:

$$L = \bigcup_{\alpha \in \mathfrak{A}_k} L(\alpha), \quad L(\alpha) \overset{\psi_\alpha}{=} \gamma_\alpha L \times P^\alpha,$$

$$\gamma_\alpha L = \bigcup_{i=1}^{\infty} (-1)^{\kappa_i(\alpha)} \beta_i S(a_1, \ldots, a_{i-1}, a_i - 1, a_{i+1}, \ldots) \tag{1.10}$$

where $\alpha = (a_1, \ldots, a_n, \ldots) \in \mathfrak{A}_k$.

We glue the manifold V with the manifold

$$-\bigcup_{\sigma \in \mathfrak{A}_{k-1}} S(\sigma) \times P^\sigma$$

along L using the identity (1.10). The $\Sigma\Gamma(k-1)$-structure on the manifold W is defined as follows:

$$-S(\sigma) \times P^\sigma \cup \left(\bigcup_{\alpha \in \xi_k^{-1}(\sigma)} V(\alpha) \right), \quad \partial W = \bigcup_{\alpha \in \xi_k^{-1}(\sigma)} M(\alpha);$$

see Figure 1.8.

The above construction defines the singular $\Sigma\Gamma(k-1)$-manifold (W, G), such that its boundary coincides with the $\Sigma\Gamma(k-1)$-manifold (M, F). So we obtain that $\gamma(k)(x) = 0$.

ii) Ker $\gamma(k) \subseteq$ Im $\partial(k)$.

Let $x = [(M, F)] \in \text{Ker } \gamma(k)$, where

$$M = \bigcup_{\alpha \in \mathfrak{A}_k} M(\alpha), \quad M(\alpha) \overset{\psi_\alpha}{\cong} \gamma_\alpha M \times P^\alpha,$$

$$F|_{M(\alpha)} : M(\alpha) \xrightarrow{\psi_\alpha} \gamma_\alpha M \times P^\alpha \xrightarrow{pr} \gamma_\alpha M \xrightarrow{f_\alpha} X.$$

The definition of the transformation $\partial(k)$ implies that the $\Sigma\Gamma(k)$-manifold (V, H) has the boundary $(\partial V, H|_{\partial V})$ such that $\partial V = M$ and $H|_{\partial V} = F$, and the identities

$$\delta V(\sigma) = \bigcup_{\alpha \in \xi^{-1}(\sigma)} M(\alpha), \quad V(\sigma) \overset{\psi_\sigma}{\cong} \gamma_\sigma V \times P^\sigma$$

are true for every $\sigma \in \mathfrak{A}_{k-1}$.

Let's note that the disjoint union

$$(L, G) = \bigcup_{\sigma \in \mathfrak{A}_{k-1}} \left(\partial V(\sigma), H|_{\partial V(\sigma)}\right)$$

is a $\Sigma\Gamma(k)$-manifold whose bordism class lies in the image of $\partial(k)$. That is, the identity

$$\partial(k) \left(\bigoplus_{\sigma \in \mathfrak{A}_{k-1}} [(\gamma_\sigma M, h_\sigma)]_\Sigma \right) = [(L, G)]_{\Sigma\Gamma(k)}$$

is true. Let $\partial V(\sigma) = \delta V(\sigma) \cup \tilde{\partial} V(\sigma)$. We consider the disjoint union of the cylinders

$$\bigcup_{\sigma \in \mathfrak{A}_{k-1}} \left(\tilde{\partial} V(\sigma) \times I\right),$$

whose top sides are glued according to the identification between the parts of the boundaries $\tilde{\partial} V(\sigma)$, $\sigma \in \mathfrak{A}_{k-1}$, of the $\Sigma\Gamma(k-1)$-manifold

V. Indeed we glue the obtained manifold U to the cylinder $L \times I$, by identifying the manifolds

$$\tilde{\partial} V(\sigma) \times \{0\} \subseteq \partial V, \quad \text{and} \quad \tilde{\partial} V(\sigma) \subset L \times \{1\}$$

for every $\sigma \in \mathfrak{A}_{k-1}$. The manifold W is a $\Sigma\Gamma(k)$-bordism between the singular $\Sigma\Gamma(k)$-manifolds M and L, as is verified directly. So we have proved that $\operatorname{Ker} \gamma(k) \subset \operatorname{Im} \partial(k)$.

The identity $\pi(k) \circ \partial(k) = \beta(k)$ follows straight from the definitions. Theorem 1.4.2 is proved. □

1.6 Σ-singularities spectral sequence

Let's construct the Σ-*singularities spectral sequence* for every locally-finite sequence Σ of closed manifolds as above. We take a pair of spaces (X, Y) and introduce the notations

$$\mathrm{D}_1^{0,t} = MG_t(X,Y), \quad \mathrm{D}_1^{s,t} = MG_t^{\Sigma\Gamma(s)}(X,Y), \quad \mathrm{E}_1^{s,t} = MG_t^{\Sigma(s)}(X,Y).$$

The transformations $\pi(k)$, $\partial(k)$, $\gamma(k)$ induce the homomorphisms

$$\mathrm{I}^{s,t} : \mathrm{D}_1^{s,t} \xrightarrow{\gamma(s)} \mathrm{D}_1^{s-1,t},$$

$$\mathrm{J}^{s,t} : \mathrm{D}_1^{s,t} \xrightarrow{\pi(s)} \mathrm{E}_1^{s+1,t+s},$$

$$\mathrm{K}^{s,t} : \mathrm{E}_1^{s,t} \xrightarrow{\partial(s)} \mathrm{D}_1^{s,t-s-1}.$$

So we have the exact couple

$$\begin{array}{ccc} \mathrm{D}_1^{*,*} & \xleftarrow{\mathrm{I}^{*,*}} & \mathrm{D}_1^{*,*} \\ & {\scriptstyle \mathrm{J}^{*,*}} \searrow \quad \nearrow {\scriptstyle \mathrm{K}^{*,*}} & \\ & \mathrm{E}_1^{*,*} & \end{array} \tag{1.11}$$

where the maps $\mathrm{I}^{*,*}$, $\mathrm{J}^{*,*}$, $\mathrm{K}^{*,*}$ have the bidegrees $(0,-1)$, $(1,s)$, $(0,-s-1)$ respectively.

Definition 1.6.1 *The spectral sequence associated with the exact couple* (1.11) *will be called the* Σ*-singularities spectral sequence* (Σ*-SSS).* □

Let's consider the filtration which induces this spectral sequence:

$$\mathsf{F}^k_*(X,Y) = \operatorname{Im}\left(MG^{\Sigma\Gamma(k)}_*(X,Y) \xrightarrow{\gamma(1)\circ\ldots\circ\gamma(k)} MG_*(X,Y)\right),$$

$$MG_*(X,Y) = \mathsf{F}^0_*(X,Y) \supset \mathsf{F}^1_*(X,Y) \supset \ldots \supset \mathsf{F}^k_*(X,Y) \supset \ldots \quad (1.12)$$

Lemma 1.6.2 *The identity*

$$\bigcap_{k\geq 1} \mathsf{F}^k_*(X,Y) = 0$$

is true for every finite dimensional CW*-complex pair* (X,Y).

Proof. We denote

$$\tilde{\mathsf{F}}_n = \bigcap_{k\geq 1} \mathsf{F}^k_n(X,Y)$$

Then we fix n and prove that $\tilde{\mathsf{F}}_n = 0$. The locally finite property of the sequence Σ implies that

$$MG^{\Sigma\Gamma(k)}_n = 0$$

for some sufficiently large number k. The finiteness of the dimension of the CW-pair (X,Y) means that $H_j(X,Y;\mathbf{Z}) = 0$ for some sufficiently large number j. The Atiyah-Hirzebruch spectral sequence

$$H_*\left(X,Y;MG^{\Sigma\Gamma(k)}_*\right) => MG^{\Sigma\Gamma(k)}_*(X,Y)$$

instantly gives that $MG^{\Sigma\Gamma(k)}_m(X,Y) = 0$ for $m > \max\{k,j\}$. It follows that $\tilde{\mathsf{F}}_n = 0$. According to Cartan and Eilenberg [31] it follows that the Σ-singularities spectral sequence converges. □

The differential d_1 of the spectral sequence coincides with the transformation

$$\beta(n) : MG^{\Sigma(n)}_*(\cdot) \longrightarrow MG^{\Sigma(n+1)}_*(\cdot),$$

which is a sum of the Bockstein operators β_k, $k = 1, 2, \ldots$, according to Lemma 1.3.2.

The other differentials

$$d_r^s \colon \mathrm{E}_r^{s,*} \longrightarrow \mathrm{E}_r^{s+r,*}$$

are the *highest Bockstein homology operations of the r-th order in the homology theory* $MG_*^{\Sigma}(\cdot)$ (their algebraic definition was given by Maunder [62]; see also [81], [124]).

The following theorem is the final result of this section.

Theorem 1.6.3 *The Σ-singularities spectral sequence has the properties:*

(i) *it is natural on the category of pairs of spaces;*

(ii) *it converges when (X, Y) is a finite dimensional CW-pair; if $\Sigma = (P)$, where $[P] = p$, then it converges to the p-prime component of the group $MG_*(X, Y)$; if $\Sigma = (P_1, \ldots, P_k, \ldots)$ and $\dim P_k > 0$, $k = 1, 2, \ldots$, then it converges to the group $MG_*(X, Y)$;*

(iii) *the differentials d_r of Σ-SSS are prime Bockstein operators (for $r = 1$) and higher order ones (for $k \geq 2$) in the theory $MG_*^{\Sigma}(\cdot)$.* □

Now let's consider the case $\Sigma = (P)$.

We denote a projection of the element $\theta = [P]$ into the term $E_1^{*,*}$ by θ. It is clear that the element θ has a degree $(1, p+1)$, where $p = \dim P$. We have the following isomorphism:

$$\mathrm{E}_1^{s,t} = MG_t^{\Sigma(s)}(X, Y) = MG_{t+s(p+1)}^{\Sigma}(X, Y).$$

So the line $E_1^{s,*}$ is the one-dimensional $MG_*^{\Sigma}(X, Y)$-module with the generator θ^s (for $s \geq 1$). It is clear that the line $E_2^{s,*}$ coincides with the s-th homology group of the following complex:

$$MG_*^{\Sigma}(X, Y) \xrightarrow{\beta} MG_*^{\Sigma}(X, Y) \to \cdots \to MG_*^{\Sigma}(X, Y) \xrightarrow{\beta} \cdots$$

So we obtain that $E_2^{0,*} = \mathrm{Ker}\ \beta$ and the line $E_2^{s,*}$ coincides with the factor

$$\mathrm{Ker}\ \beta / \mathrm{Im}\ \beta$$

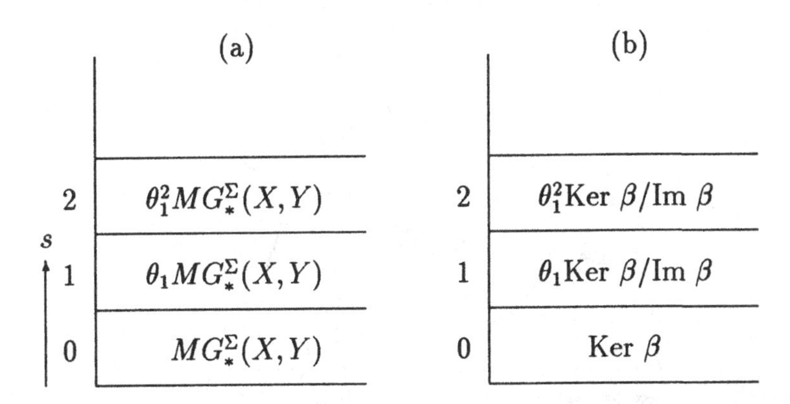

Figure 1.9: Terms (a) $E_1^{s,*}$, (b) $E_2^{s,*}$.

with the generator θ^s for every $s \geq 1$; see Figure 1.9.

A geometric interpretation of the spectral sequence is evident. That is, we resolve the singularities (see Figure 1.10, where we show the procedure for the element $x = [M]_\Sigma$ from the zero line).

Of course, the general case is much more complicated. Let $\Sigma = (P_1, \ldots, P_k, \ldots)$. We denote the element $[P_k]$ as well as its projection into the term $E_1^{*,*}$ of the spectral sequence, by θ_k for every $k = 1, 2, \ldots$. Their degrees are equal to $(1, p_k + 1)$ where $p_k = \dim P_k$.

The line $E_1^{s,*}$ coincides with the $MG_*^\Sigma(X,Y)$-module with the generators

$$\theta_{i_1}^{a_1} \theta_{i_2}^{a_2} \ldots \theta_{i_m}^{a_m},$$

for $s \geq 1$, where $a_1 + \ldots + a_m = s$, $a_j \geq 0$.

The line $E_1^{s,*}$ is the s-th homology group of the complex

$$MG_*^{\Sigma(1)}(X,Y) \xrightarrow{\beta(1)} MG_*^{\Sigma(2)}(X,Y) \to \ldots \to MG_*^{\Sigma(k)}(X,Y) \xrightarrow{\beta(k)} \ldots$$

So a structure of the term $E_2^{*,*}$ depends on the action of the Bockstein operators β_k on $MG_*^\Sigma(X,Y)$.

Figure 1.11 shows what happens with the element $x = [M]_\Sigma$ from the zero line (in the case when $\Sigma = (P_1, P_2)$).

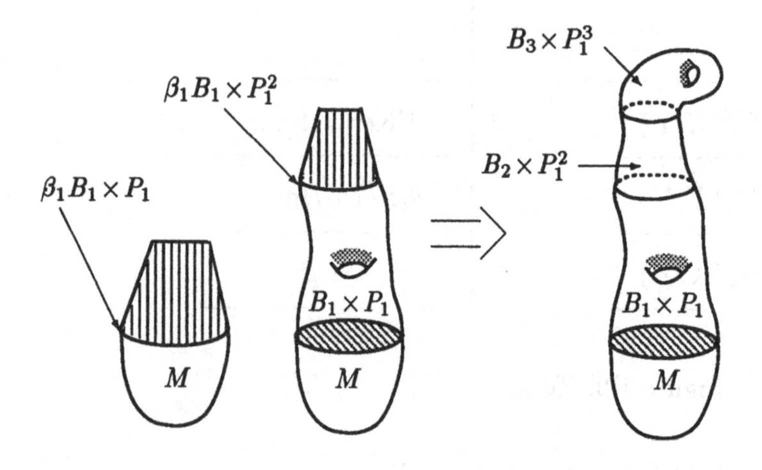

Figure 1.10: Geometric meaning of Σ-*SSS* for one singularity.

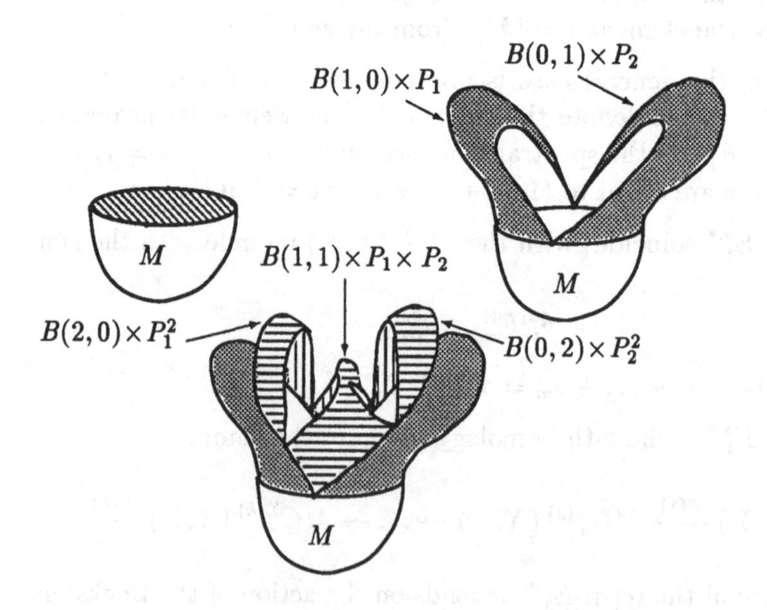

Figure 1.11: Geometric sence of Σ-*SSS* for two singularities.

Note 1.6.1 *The exact triangles (1.9) induce the following diagram of representing spectra:*

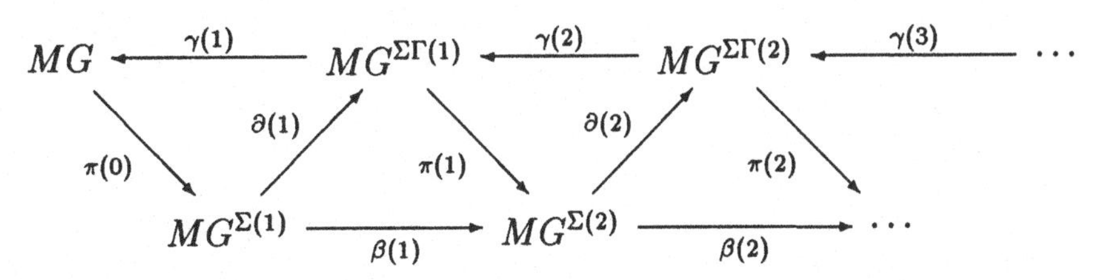

The above diagram presents also the Adams resolution of the spectrum MG in the theory $MG^*_\Sigma(\cdot)$ and the corresponding Σ-singularities spectral sequence coincides with the Adams-Novikov spectral sequence. We will deal with the subject in Chapter 3. □

Chapter 2

Product structures

The Σ-singularities spectral sequence has provided us with the geometric procedure for restoring the bordism theory $MG_*(\cdot)$ out of the bordism theory with singularities $MG_*^{\Sigma}(\cdot)$. To convert the construction into a computation tool we need to have a natural multiplicative structure in this spectral sequence. Even in the simplest cases we can't have any hope of doing computations successfully without a product structure.

The spectral sequence has a product structure of this kind only when the corresponding bordism theory $MG_*^{\Sigma}(\cdot)$ has some *admissible multiplicative structure*, i.e. a structure which is compatible with the ordinary product structure in the bordism theory $MG_*(\cdot)$. If we have only one singularity then a product structure in the Σ-singularity spectral sequence can be determined by a direct geometric construction; see section 2.6. The problem becomes too complicated in the case of many singularities. Anyway we can do without direct construction. That is, we can identify the Σ-singularities spectral sequence with the corresponding Adams-Novikov spectral sequence and then make use of the resulting multiplicative properties. Both ways give us the desired product structure determined by a geometric product structure in the bordism theory with singularities.

O.K.Mironov has described a geometric construction of the *admissible product structure* in bordism theories with singularities; see [67],

[68]. We follow some of his ideas here as well as the present author's paper [16]; but as was mentioned above the most general case is not the subject of our consideration; rather we are going to dwell upon a restricted situation sufficient for our purposes.

The fact is that the construction does work well when the dimensions of the manifold-singularities are even or their bordism classes are of order two. Other restrictions besides those mentioned above are not assumed. We hope that the basic construction of the admissible product structure is rather simple and obvious.

2.1 Multiplicative structures

Here the problem of interest is when the bordism theory with singularities $MG^{\Sigma}_*(\cdot)$ admits a product structure compatible with the ordinary one in the theory $MG_*(\cdot)$. Let us recall some definitions and constructions.

Definition 2.1.1 *The collection of the maps*

$$\mu_{(X,X_1;Y,Y_1)} : h_*(X,X_1) \otimes h_*(Y,Y_1) \longrightarrow h_*(X \times Y, X_1 \times Y \cup X \times Y_1)$$

is called an external product structure in the homology theory $h_*(\cdot)$ if they are determined for all pairs of spaces $(X,X_1),(Y,Y_1)$ and satisfy the following axioms.

1° The maps $\mu_{(X,X_1;Y,Y_1)}$ are natural with respect to the maps of the pairs of spaces.

2° The maps $\mu_{(X,X_1;Y,Y_1)}$ are compatible with the boundary homomorphisms.

3° There exists a two-sided unit $\mathbf{1} \in \tilde{h}(S^0)$.

These are minimal requirements which a product structure should satisfy. The most interesting is the case when the product structure μ also satisfies the following axioms 4°, 5°.

4° **Commutativity.** *The diagram*

$$\begin{array}{ccc} h_*(X)\otimes h_*(Y) & \xrightarrow{\mu} & h_*(X\times Y) \\ {\scriptstyle I}\downarrow & & \downarrow{\scriptstyle \chi_*} \\ h_*(X)\otimes h_*(Y) & \xrightarrow{\mu} & h_*(X\times Y) \end{array} \tag{2.1}$$

is commutative for all spaces X, Y , *where*

$$\chi : X \times Y \longrightarrow X \times Y$$

is the permutation of the factors, and I *is the following canonical isomorphism:*

$$I(x\otimes y) = (-1)^{\deg x \deg y} y \otimes x.$$

5° **Associativity.** *The following diagram is commutative for all* X, Y, Z :

$$\begin{array}{ccc} h_*(X)\otimes h_*(Y)\otimes h_*(Z) & \longrightarrow & h_*(X\times Y)\otimes h_*(Z) \\ \downarrow & & \downarrow \\ h_*(X)\otimes h_*(Y\times Z) & \longrightarrow & h_*(X\times Y\times Z) \end{array} \tag{2.2}$$

A direct product of the manifolds induces the external product structure in the bordism theories $MG_*(\cdot)$ which satisfies the axioms 1° — 5°.

The bordism theory with singularities $MG_*^{\Sigma}(\cdot)$ naturally possesses a *module structure over the bordism theory* $MG_*(\cdot)$. This means that there exist the bordism theory *pairings*

$$\mu_R : MG_*^{\Sigma}(\cdot)\otimes MG_*(\cdot) \longrightarrow MG_*^{\Sigma}(\cdot),$$

$$\mu_L : MG_*(\cdot)\otimes MG_*^{\Sigma}(\cdot) \longrightarrow MG_*^{\Sigma}(\cdot).$$

Definition 2.1.2 *The external product* μ *in the bordism theory* $MG_*^{\Sigma}(\cdot)$ *is called an admissible product structure when it satisfies the axioms* 1° - 3° *and is compatible with the pairings* μ_L, μ_R , *i.e. the following diagram is commutative:*

$$\begin{array}{ccc}
MG^{\Sigma}_*(\cdot)\otimes MG_*(\cdot) & \xrightarrow{\mu_R} & MG^{\Sigma}_*(\cdot) \\
\downarrow & & \downarrow \\
MG^{\Sigma}_*(\cdot)\otimes MG^{\Sigma}_*(\cdot) & \xrightarrow{\mu} & MG^{\Sigma}_*(\cdot) \\
\uparrow & & \uparrow \\
MG_*(\cdot)\otimes MG^{\Sigma}_*(\cdot) & \xrightarrow{\mu_L} & MG^{\Sigma}_*(\cdot)
\end{array} \tag{2.3}$$

As was already mentioned it is convenient to deal with manifolds whose cones over singularities are *omitted*, i.e. with Σ-manifolds. It is to be remembered that all constructions should be compatible with the projections on the models of Σ-manifolds.

A geometric procedure to determine a product structure on the bordism theory $MG^{\Sigma}_*(\cdot)$ may be divided into the following steps. Firstly, we construct the *product* $M \cdot N$ for all Σ-manifolds M, N, i.e. a Σ-manifold which is bordant to the direct product $M \times N$ when one of the factors has empty set of singularities. Secondly we extend the construction up to singular Σ-manifolds and verify its compatibility with the boundary operator.

The above procedure give us a well defined product structure on bordism classes. To formalize this technique we would like to apply a notion of *canonical constructions* which was introduced by O.K.Mironov [68].

Let us take the original bordism theory $MG_*(\cdot)$ and denote the corresponding category of manifolds by $\mathfrak{M}$. As was mentioned above the sequence $\Sigma = (P_1, \ldots, P_k, \ldots)$ of the closed manifolds is supposed to be locally-finite. Let's consider the sequences $\Sigma_k = (P_1, \ldots, P_k)$ for all $k = 1, 2, \ldots$ too. Let $\mathfrak{M}_k$ denote a corresponding category of Σ_k-manifolds.

Definition 2.1.3 *A canonical n-linear construction $\mathfrak{A}$ of dimension p is a rule which determines the Σ-manifold $\mathfrak{A}(M_1, \ldots, M_n)$ of dimension $(m_1 + \ldots + m_n + p)$, the submanifold $\mathfrak{B}(M_1, \ldots, M_n)$ of its boundary $\delta\mathfrak{A}(M_1, \ldots, M_n)$ of dimension $(m_1+\ldots+m_n+p-1)$ and the continuous* *maps*

$$\pi(M_1,\ldots,M_n) : \mathfrak{A}(M_1,\ldots,M_n) \longrightarrow (M_1)_\Sigma \times \ldots \times (M_n)_\Sigma$$

for every collection $(M_1^{m_1},\ldots,M_n^{m_n})$ *of* Σ*-manifolds.*

The rule should have the following properties:

1. Compatibility with the boundary operator. There exists the decomposition

$$\delta\mathfrak{A}(M_1,\ldots,M_n) = \mathfrak{B}(M_1,\ldots,M_n) \cup \left(\bigcup_{I\subset\{1,\ldots,n\}} \mathfrak{A}\left(\delta_I(M_1,\ldots,M_n)\right)\right),$$

where the map δ_I converts the collection $(M_1,\ldots,M_n)$ into the collection $(L_1,\ldots,L_n)$, which is defined as follows:

$$L_j = \left\{ \begin{array}{ll} M_j & \text{if } \ j \notin I, \\ \delta M_j & \text{if } \ j \in I. \end{array} \right\} \tag{2.4}$$

The decomposition has to satisfy the following identities:

$$\mathfrak{B}(M_1,\ldots,M_n) \cap \mathfrak{A}\left(\delta_I(M_1,\ldots,M_n)\right) = \mathfrak{B}\left(\delta_I(M_1,\ldots,M_n)\right),$$

$$\delta\mathfrak{A}\left(\delta_I(M_1,\ldots,M_n)\right) = \mathfrak{B}\left(\delta_I(M_1,\ldots,M_n)\right) \cup \left(\bigcup_{J\subset I} \mathfrak{B}\left(\delta_J(M_1,\ldots,M_n)\right)\right),$$

$$\pi(M_1,\ldots,M_n)|_{\mathfrak{A}(\delta_I(M_1,\ldots,M_n))} = \pi\left(\delta_I(M_1,\ldots,M_n)\right).$$

2. Compatibility with the direct product on the manifold L^l without singularities. The following identities should be true:

$$\mathfrak{A}(M_1,\ldots,M_{j-1},M_j\times L,M_{j+1},\ldots,M_n) =$$
$$(-1)^{l(m_{j+1}+\ldots+m_n)}\mathfrak{A}(M_1,\ldots,M_n)\times L,$$
$$\mathfrak{B}(M_1,\ldots,M_{j-1},M_j\times L,M_{j+1},\ldots,M_n) =$$
$$(-1)^{l(m_{j+1}+\ldots+m_n)}\mathfrak{B}(M_1,\ldots,M_n)\times L,$$
$$\pi(M_1,\ldots,M_{j-1},M_j\times L,M_{j+1},\ldots,M_n) = \pi(M_1,\ldots,M_n)\times \mathrm{Id}.$$

3. Compatibility with a gluing. If the manifold M has the decomposition

$$M_j = M_j' \cup_{M'''} M_j''$$

then the following identities should be satisfied:

$$\mathfrak{A}(M_1, \dots, M_j, \dots, M_n) =$$
$$\mathfrak{A}(M_1, \dots, M_j', \dots, M_n) \cup_{\mathfrak{A}(M_1, \dots, M_j''', \dots, M_n)} \mathfrak{A}(M_1, \dots, M_j'', \dots, M_n),$$

$$\mathfrak{B}(M_1, \dots, M_j, \dots, M_n) =$$
$$\mathfrak{B}(M_1, \dots, M_j', \dots, M_n) \cup_{\mathfrak{B}(M_1, \dots, M_j''', \dots, M_n)} \mathfrak{B}(M_1, \dots, M_j'', \dots, M_n),$$

$$\pi(M_1, \dots, M_j, \dots, M_n) =$$
$$\pi(M_1, \dots, M_j', \dots, M_n) \cup_{\pi(M_1, \dots, M_j''', \dots, M_n)} \pi(M_1, \dots, M_j'', \dots, M_n).$$

□

A canonical construction

$$\partial\mathfrak{A} = (\mathfrak{B}, \emptyset, \pi|_{\mathfrak{B}})$$

is called a *boundary of the canonical construction* $\mathfrak{A}$. The canonical construction $\mathfrak{A}$ is *closed* if $\mathfrak{B}(M_1, \dots, M_n) = \emptyset$ for every collection $(M_1, \dots, M_n)$ of Σ-manifolds.

Simple examples of canonical constructions are the operator of multiplying by an ordinary manifold without singularities $M \to M \times N$ and the Bockstein operator $\beta_k : M \to \beta_k M$.

Canonical constructions have the following properties.

1) The result of gluing along the boundaries is also a canonical construction.

2) A composition of canonical constructions is also a canonical construction.

The definitions mentioned above imply the following.

Lemma 2.1.4 *A closed n-linear canonical construction $\mathfrak{A}$ of dimension p determines the following natural MG_*-module homomorphisms:*

$$MG^{\Sigma}_*(\mathfrak{A}) : MG^{\Sigma}_{m_1}(X_1) \otimes \ldots \otimes MG^{\Sigma}_{m_n}(X_n) \longrightarrow MG^{\Sigma}_q(X_1 \times \ldots \times X_n),$$

which commute with the boundary operators for every collection

$$(X_1, \ldots, X_n)$$

of CW-complexes, where $q = m_1 + \ldots + m_n$. If the canonical construction $\mathfrak{A}$ is a boundary of some other one then the homomorphisms $MG^{\Sigma}_*(\mathfrak{A})$ must be trivial for every collection $(X_1, \ldots, X_n)$. □

So to determine the product structure in the bordism theory $MG^{\Sigma}_*(\cdot)$ it is sufficient to obtain a corresponding canonical construction. The following lemma is also implied by the above definitions.

Lemma 2.1.5 *Let $\mathfrak{A}$ be a canonical bilinear construction of zero dimension on the category of Σ-manifolds. Then if $\mathfrak{A}(pt, pt) = pt$ it determines the admissible product structure in the theory $MG^{\Sigma}_*(\cdot)$.* □

2.2 Existence of product structure

Suppose $MG^{\Sigma}_*(\cdot)$ is the bordism theory where the sequence of the closed manifolds $\Sigma = (P_1, \ldots, P_k, \ldots)$ is the locally-finite one as before. Let $\Sigma_k = (P_1, \ldots, P_k)$ be the first part of the sequence for $k = 1, 2, \ldots$ and let the whole sequence Σ satisfy the following condition:

(*) $2[P_k] = 0$ *in the group $MG^{\Sigma_{k-1}}_*$ if the dimension of the manifold P_k is odd.*

The condition (*) is sufficient for our purpose.

Let's consider the manifold

$$P'_k = P^{(1)}_k \times P^{(2)}_k \times I,$$

where $P_k^{(1)}$, $P_k^{(2)}$ are copies of the manifold P_k, as a Σ_k-manifold in the following way:

$$\partial_k P'_k = \partial P'_k = \beta_k P'_k \times P_k,$$

$$\beta_k P'_k = (-1)^{p_k+1} P_k^{(1)} \times \{0\} \cup P_k^{(2)} \times \{1\}, \quad p_k = \dim P_k.$$

Note that the Σ_k-manifold P'_k is bordant to a manifold without singularities under the condition (*).

The Σ_k-manifold P'_k is an obstruction to the existence of the admissible product structure μ_k in the bordism theory $MG_*^{\Sigma_k}(\cdot)$ for the following reasons.

Let M be some closed Σ_k-manifold. The diagram (2.3) implies that

$$\mu_k\left([M] \otimes [P_k]\right) = 0.$$

The manifold P'_k has no Σ_i-singularities for every $i < k$, so the direct product $\beta_k M \times P'_k$ has a well determined Σ_k-manifold structure.

Lemma 2.2.1 *The identity*

$$[M \times P_k]_{\Sigma_k} = [\beta_k M \times P'_k]_{\Sigma_k}$$

is true in the group $MG_*^{\Sigma_k}$ for every closed Σ_k-manifold M.

Note 2.2.1 *There is another Σ_k-structure on the manifold $M \times P_k$ which is induced by the Σ_k-structure on the manifold M. The projection of the manifold $M \times P_k$ onto its direct factor M restricted to the boundary*

$$\partial_k M \times P_k = \beta_k M \times P_k^{(1)} \times P_k^{(2)}$$

results in a Σ_k-structure on $M \times P_k$, which doesn't coincide with the above one.

Proof of Lemma 2.1.1. Let's consider the cylinder

$$M \times P_k \times [0, 2] = C$$

as a Σ_k-manifold having the boundary (see Figure 2.1)

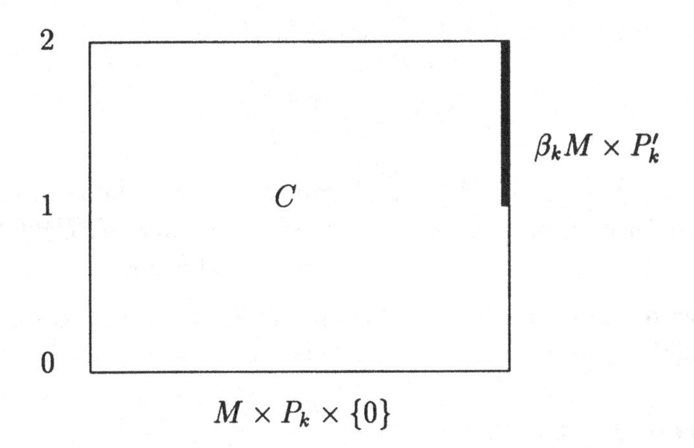

Figure 2.1: Σ_k-manifold $M \times P_k \times [0,2] = C$.

$$\delta C = M \times P_k \times \{0\} \cup \beta_k M \times P_k^{(1)} \times P_k^{(2)} \times [1,2]$$

$$= M \times P_k \times \{0\} \cup \beta_k M \times P'_k.$$

Also we have

$$\partial_j C = \partial_j M \times P_k \times [0,2], \quad j < k,$$

$$\partial_k M = -M \times P_k \times \{2\} \cup \beta_k M \times P_k^{(1)} \times P_k^{(2)} \times [0,1]$$

$$= -M \times P_k \times \{2\} \cup (-1)^{p_k+1} \left(\beta_k M \times P_k^{(1)} \times [0,1]\right) \times P_k^{(2)}. \quad \square$$

Let us consider the simplest example which illustrates the situation. We introduce the singularity $\Sigma_0 = (P_0)$ into the bordism theory $MSU_*(\cdot)$ where $[P_0] = q$. It is simple to verify that

$$[P'_0]_{\Sigma_0} = \frac{q(q+1)}{2}\theta_1,$$

where the element $\theta_1 = [P_1]$ is a generator of the group $MSU_1 = \mathbf{Z}/2$. Let's consider an SU-manifold W such that $\partial W = 2P_1$ as a Σ_0-manifold by assuming that $\beta W = P_1$. Lemma 2.1.1 implies that

$$q\,[W]_{\Sigma_0} = \frac{q(q+1)}{2}\theta_1^2.$$

In particular, we get $q\,[W] \neq 0$ when $q = 4k+2$, so an admissible product structure doesn't exist in the bordism theory $MSU_*^{\Sigma_0}(\cdot)$. Similar examples may be given for the symplectic and framed bordism theories.

Now let us formulate a sufficient condition for the existence of the admissible product structure in the bordism theory $MG_*^{\Sigma_k}(\cdot)$.

Theorem 2.2.2 *Let the equality*

$$[P_i']_{\Sigma_i} = 0$$

be true for every $i = 1,\ldots,k$, where Q_i are Σ_i-manifolds such that $\delta Q_i = P_i'$. Then there exists the admissible product structure $\mu_i = \mu_i(Q_1,\ldots,Q_k)$ in the bordism theory $MG_*^{\Sigma_i}(\cdot)$ for every i which makes the following diagram commutative:

$$\begin{array}{ccc} MG_*^{\Sigma_{i-1}}(\cdot)\otimes MG_*^{\Sigma_{i-1}}(\cdot) & \xrightarrow{\mu_{i-1}} & MG_*^{\Sigma_{i-1}}(\cdot) \\ \downarrow{\scriptstyle \pi_i^{i-1}\otimes\pi_i^{i-1}} & & \downarrow{\scriptstyle \pi_i^{i-1}} \\ MG_*^{\Sigma_i}(\cdot)\otimes MG_*^{\Sigma_i}(\cdot) & \xrightarrow{\mu_i} & MG_*^{\Sigma_i}(\cdot) \end{array} \qquad (2.5)$$

Here $MG_*^{\Sigma_0}(\cdot) = MG_*(\cdot)$ and μ_0 is the original product structure in the bordism theory $MG_*(\cdot)$.

Proof. Every object of the category $\mathfrak{M}_k$ is well determined up to a diffeomorphism preserving the G-structure. So we can suppose that every Σ_k-manifold M considered has a collar along its boundary ∂M. Let's examine the case when $k = 1$. The direct product $M \times N$ of Σ_1-manifolds has the following decomposition of its boundary:

$$\partial(M\times N) = \left(\beta_1 M \times P_1^{(1)} \times N\right) \cup \left(\beta_1 M \times P_1^{(1)} \times \beta_1 N \times P_1^{(2)} \times I\right)$$
$$\cup\, (-1)^m \left(M \times \beta_1 N \times P_1^{(2)}\right).$$

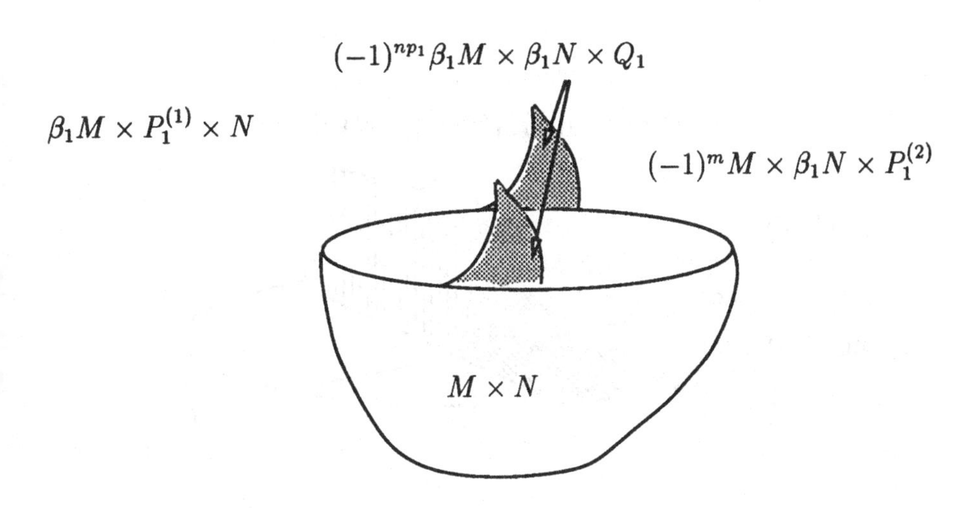

Figure 2.2: Σ_1-manifold $\mathfrak{m}_1(M, N)$.

We note that the manifold

$$\beta_1 M \times P_1^{(1)} \times \beta_1 N \times P_1^{(2)} \times I$$

is diffeomorphic to the manifold

$$(-1)^{np_1}\beta_1 M \times \beta_1 N \times P_1',$$

which can be bounded by the manifold

$$(-1)^{np_1}\beta_1 M \times \beta_1 N \times Q_1.$$

So the pairing

$$\mathfrak{m}_1 : \mathfrak{M}_1 \times \mathfrak{M}_1 \longrightarrow \mathfrak{M}_1$$

is well defined; see Figure 2.2. To finish the induction now suppose that the bilinear canonical construction $\mathfrak{m}_i$ is determined for every $i = 1, \ldots, k-1$, such that its boundary is equal to the Σ_i-manifold

$$\delta\mathfrak{m}_i(M^m, N^n) = \mathfrak{m}_i(\delta M^m, N^n) \bigcup_{\mathfrak{m}_i(\delta M^m, \delta N^n)} \mathfrak{m}_i(M^m, \delta N^n).$$

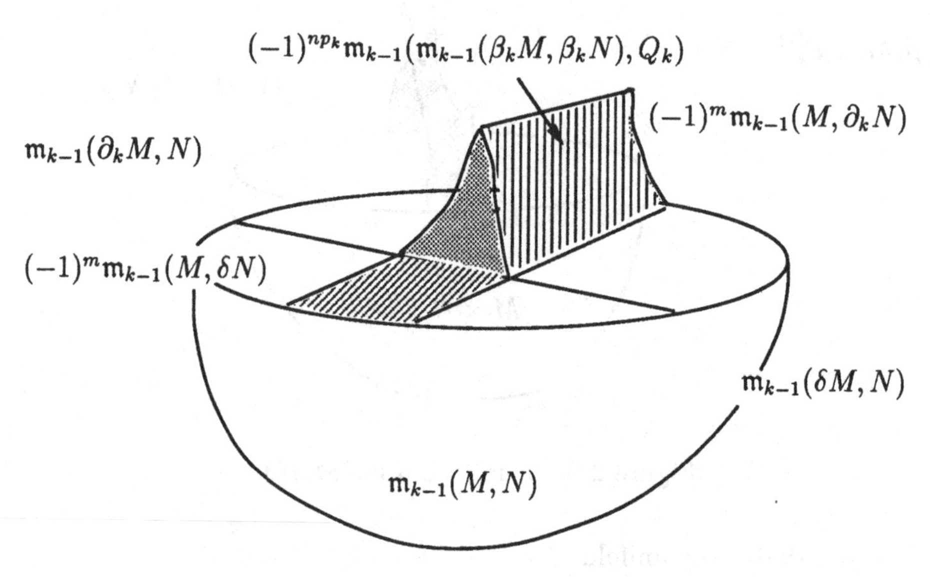

Figure 2.3: Σ_k-manifold $\mathfrak{m}_k(M, N)$.

Let's consider the Σ_k-manifolds M^m and N^n; the boundary of the Σ_{k-1}-manifold $\mathfrak{m}_{k-1}(M^m, N^n)$ has the following decomposition:

$$\delta\mathfrak{m}_{k-1}(M^m, N^n) = \mathfrak{m}_{k-1}(\delta M^m, N^n) \cup (\mathfrak{m}_{k-1}(\delta M^m, \delta N^n) \times I)$$
$$\cup (-1)^m \mathfrak{m}_{k-1}(M^m, \delta N^n) \cup \mathfrak{m}_{k-1}(\partial_k M^m, N^n)$$
$$\cup (\mathfrak{m}_{k-1}(\partial_k M^m, \partial_k N^n) \times I) \cup (-1)^m \mathfrak{m}_{k-1}(M^m, \partial_k N^n).$$

Definition 2.1.3 implies that the bilinear canonical construction $\mathfrak{m}_{k-1}$ satisfies the following equality:

$$\mathfrak{m}_{k-1}(M^m, \partial_k N^n) \times I = (-1)^{np_k}\mathfrak{m}_{k-1}(\beta_k M^m, \beta_k N^n) \times P'_k.$$

The bilinear canonical construction $\mathfrak{m}_k$ is determined in the following way:

$$\mathfrak{m}_k(M^m, N^n) = \mathfrak{m}_{k-1}(M^m, N^n)$$

$$\cup (-1)^{np_k}\mathfrak{m}_{k-1}(\mathfrak{m}_{k-1}(\beta_k M^m, \beta_k N^n), Q_k);$$

see Figure 2.3. Verification of the properties 1-3 of Definition 2.1.3 may be done directly. We note also that $\mathfrak{m}_k(M^m, N^n) = \mathfrak{m}_{k-1}(M^m, N^n)$ for all Σ_{k-1}-manifolds M, N. It is also clear that $\mathfrak{m}_k(pt, pt) = pt$; so the canonical construction $\mathfrak{m}_k$ determines the admissible product structure $\mu_k = \mu_k(Q_1, \ldots, Q_k)$ in the bordism theory $MG_*^{\Sigma_k}(\cdot)$. The diagram (2.5) is commutative according to the above constructions. □

The product properties of the Bockstein operators $\beta_k, k = 1, 2, \ldots$, are very important for our purposes. To describe them we should more carefully consider the manifolds Q_k bounding the obstructions P_k'. The manifold $\beta_k Q_k$ is a Σ_{k-1}-manifold with the boundary

$$\delta \beta_k Q_k = (-1)^{p_k+1} P_k \cup P_k.$$

Let's define the structure of a Σ_k-manifold on $\beta_k Q_k$ by putting

$$\partial_j(\beta_k Q_k) = \left\{ \begin{array}{ll} \partial_j(\beta_k Q_k) & \text{if } \ j \neq k, \\ (-1)^{p_k+1} P_k^{(1)} \cup P_k^{(2)} & \text{if } \ j = k. \end{array} \right\} \tag{2.6}$$

We denote this Σ_k-manifold by $\widehat{\beta_k} Q_k$. The following formula may be proved by simple induction:

$$\beta_k \mathfrak{m}_k(M^m, N^n) = (-1)^{p_k n} \mathfrak{m}_k(\beta_k M^m, N^n) \tag{2.7}$$

$$\cup \, (-1)^{p_k n + n + m} \mathfrak{m}_k(\mathfrak{m}_k(\beta_k M, \beta_k N), \widehat{\beta_k} Q_k) \cup (-1)^m \mathfrak{m}_k(M, \beta_k N).$$

The corresponding product formula for Bockstein operators β_i is much more complicated when $i < k$. It depends upon the Σ_j-singularities of the Σ_n-manifolds Q_n for $j < n$.

Anyway there exists an important case when these formulas are quite simple. We denote by w_i the element

$$-\left[\widehat{\beta_i} Q_i\right]_{\Sigma_i}$$

for every $i = 1, \ldots, k$. The following lemma may be proved by simple induction.

Lemma 2.2.3 *Let the Σ_i-manifolds Q_i be such that*

$$[\beta_i Q_i]_{\Sigma_k} = 0$$

for all $i \neq j$. Then the formula

$$\begin{aligned} \beta_i\left(\mu_k(x\otimes y)\right) &= (-1)^{p_i n}\mu_k(\beta_i x\otimes y)+(-1)^m\mu_k(x\otimes\beta_i y) \\ &+(-1)^{p_i n}\mu_k\left(\mu_k(\beta_i x\otimes\beta_i y)\otimes w_i\right) \end{aligned} \tag{2.8}$$

holds for every $i = 1, 2, \ldots$ where $x \in MG_*^{\Sigma_k}(X, X_1)$, $y \in MG_*^{\Sigma_k}(Y, Y_1)$. □

The dependence of the product structure μ_k upon the Σ_i-manifolds Q_i can also be examined for $i = 1, \ldots, k-1$. Let us suppose that two different Σ_i-manifolds $Q_k^{(1)}, Q_k^{(2)}$ bound the obstruction P_k'. We define the element

$$c_k = \left[-Q_k^{(1)} \cup Q_k^{(2)}\right]_{\Sigma_k}$$

of the ring $MG_*^{\Sigma_k}$. The product structure in the theory $MG_*^{\Sigma_{k-1}}(\cdot)$ is denoted by $\mu_{k-1} = \mu_{k-1}(Q_1, \ldots, Q_{k-1})$ and two admissible product structures in the bordism theory $MG_*^{\Sigma_{k-1}}(\cdot)$ by

$$\mu_k^{(1)} = \mu_k(Q_1, \ldots, Q_{k-1}, Q_k^{(1)}), \quad \mu_k^{(2)} = \mu_k(Q_1, \ldots, Q_{k-1}, Q_k^{(2)}).$$

Theorem 2.2.4 *Every two elements*

$$x \in MG_*^{\Sigma_k}(X, X_1), \qquad y \in MG_*^{\Sigma_k}(Y, Y_1)$$

satisfy the following formula:

$$\mu_k^{(1)}(x\otimes y)-\mu_k^{(2)}(x\otimes y) = (-1)^{(\deg y)p_k}\mu_{k-1}(\mu_{k-1}(\beta_k x\otimes\beta_k y)\otimes c_k). \tag{2.9}$$

Proof may be done by direct comparing of the constructions. □

Note 2.2.2 *So we can see that the admissible product structure $\mu_k = \mu_k(Q_1, \ldots, Q_k)$ can be corrected by some element of the group $MG_q^{\Sigma_k}$ where $q = \dim Q_k = 2\dim P_k + 2$. In particular if we take $Q_k^{(2)} = Q_k^{(1)} \cup S$ where S is some Σ_{k-1}-manifold then the product formula for the Bockstein operator β_k should be the same. Product properties of the operator β_i depend on the manifold S in the case $i < k$.* □

Finally we would like to formulate one possible generalization of Theorem 2.1.2 (it is proved by simple induction [16]). There can be a situation when the obstructions $[P_i']_{\Sigma_i}$ are not zero in the groups $MG_*^{\Sigma_i}$ for $i = 1, \ldots, k-1$, and at the same time they are zero in the group $MG_*^{\Sigma_k}$ as well as the obstruction $[P_k']_{\Sigma_k}$, i.e. $[P_i']_{\Sigma_k} = 0$ for $i = 1, \ldots, k$. Then an admissible product structure in the bordism theory $MG_*^{\Sigma_k}(\cdot)$ may be constructed too.

Theorem 2.2.5 *Let the sequence* $\Sigma_k = (P_1, \ldots, P_k)$ *be such that* $\dim P_i \neq \dim P_j$ *for* $i \neq j$. *Then the admissible product structure* μ_k *in the theory* $MG_*^{\Sigma_k}(\cdot)$ *exists if the obstructions*

$$[P_1']_{\Sigma_k} = 0 \ , \ \ldots \ , \ [P_k']_{\Sigma_k} = 0$$

are zero. □

For example, let us consider the bordism theory $MSU_*^{\Sigma}(\cdot)$, where $\Sigma = (P_0, P_1)$, $[P_0] = 2$, $[P_1] = \theta_1$. We also denote $\Sigma_0 = (P_0)$. As we have seen above, the bordism theory $MSU_*^{\Sigma_0}(\cdot)$ doesn't have an admissible product structure, while the theory $MSU_*^{\Sigma}(\cdot)$ does.

2.3 Commutativity for the case of one singularity

Here we consider the case when $\Sigma = (P)$. Let $[P']_{\Sigma} = 0$ and suppose the Σ-manifold Q is such that $\delta Q = P'$, which determines the admissible product structure μ in the bordism theory $MG_*^{\Sigma}(\cdot)$. The involution

$$\tau : P' \longrightarrow P'$$

on the manifold P' is determined by the formula

$$\tau(p_1, p_2, t) = (p_2, p_1, 1-t),$$

where (p_1, p_2, t) are coordinates in the product $P^{(1)} \times P^{(2)} \times I$. So we have the Σ-manifold

$$B = Q \bigcup_{P' = \tau P'} -Q.$$

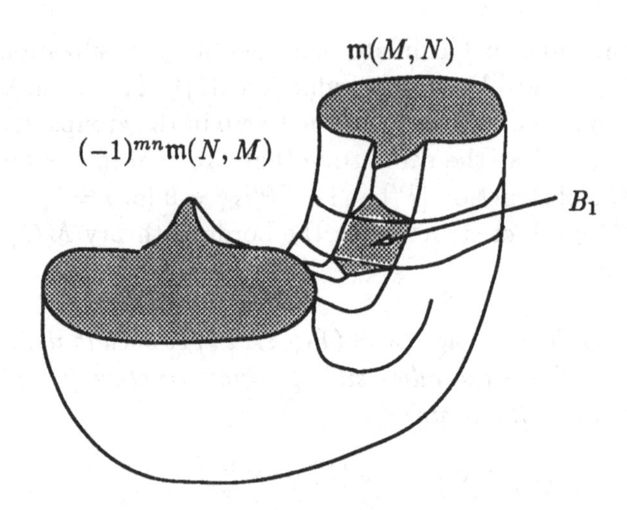

Figure 2.4: Σ-manifold $\mathfrak{B}(M,N)$.

Theorem 2.3.1 *Every two elements*

$$x \in MG_*^\Sigma(X, X_1), y \in MG_*^\Sigma(Y, Y_1)$$

satisfy the following equality:

$$\begin{aligned} &\mu(x \otimes y) - (-1)^{\dim x \dim y} \chi_* \mu(y \otimes x) \\ &= (-1)^{(p+1)\cdot \dim y + 1} \mu(\mu(\beta(x) \otimes [B]_\Sigma) \otimes \beta(y)), \end{aligned} \tag{2.10}$$

where $\chi : X \times Y \longrightarrow Y \times X$ is the factors' permutation.

Proof. Let M^m and N^n be Σ-manifolds and $\alpha = n(p+1)+1$.

Note that we have supposed the manifolds M and N have collars along their boundaries. So we have

$$\mathfrak{m}(M,N) = M \times N \cup (-1)^\alpha \beta M \times \beta N \times Q = \Delta_1,$$

$$(-1)^{mn}\mathfrak{m}(N,M) = M \times N \cup (-1)^\alpha \beta M \times \beta N \times Q = \Delta_2,$$

where in the first case

$$\beta M \times \beta N \times \delta Q = \beta M \times \beta N \times P',$$

and in the second case

$$\beta M \times \beta N \times \delta Q = \beta M \times \beta N \times \tau P'.$$

Let us consider two cylinders $\Delta_1 \times I$ and $\Delta_2 \times I$ and glue them by the cylinder $M \times N \times I_3$ as follows:

$$M \times N \times \{1\} = M \times N \times \{0\} \subset \Delta_1 \times \{0\},$$

$$M \times N \times \{0\} = M \times N \times \{1\} \subset \Delta_1 \times \{1\}.$$

The manifold obtained is denoted by $\mathfrak{B}(X, Y)$; see Figure 2.4. Its boundary is the manifold

$$\delta\mathfrak{B}(X, Y) = M \times N \cup (-1)^{\alpha} \beta M \times \beta N \times B_1 \cup (-1)^{mn} N \times M,$$

where

$$B_1 = Q^{(1)} \bigcup_{\delta Q = P' \times \{0\}} P' \times I \bigcup_{\delta Q = \tau P' \times \{1\}} -Q^{(2)}.$$

The Σ-manifold B_1 is diffeomorphic to the Σ-manifold B. So the manifold $\mathfrak{B}(X, Y)$ determines the bilinear construction $\mathfrak{B}$. According to Lemma 2.1.1 the equality (2.8) holds. □

In particular, we get that $[B]_\Sigma = 0$ in the group MG_*^Σ. So this is a sufficient condition for commutativity of the admissible product structure μ.

The following simple lemma will be very useful for applications.

Lemma 2.3.2 *If the number* $p = \dim P$ *is odd then* $2[B]_\Sigma = 0$. □

The element $b = [B]_\Sigma$ depends on the choice of the manifold Q, bounding the Σ-manifold P'. Suppose $Q^{(1)}, Q^{(2)}$ are two such manifolds, $\mu(Q^{(1)})$, $\mu(Q^{(2)})$ are the corresponding product structures, and $b^{(1)}, b^{(2)}$ are the obstructions to their commutativity. We denote the element from the group MG_*^Σ by c, which is determined by the Σ-manifold

$$Q^{(1)} \bigcup_{P'} -Q^{(2)}.$$

We have the next equality according to the definitions of the elements $b^{(1)}$, $b^{(2)}$, c:

$$b^{(1)} - b^{(2)} = 2c.$$

So the element

$$k(P) \equiv b \mod 2 \tag{2.11}$$

of the group $MG_*^\Sigma \otimes \mathbf{Z}/2$ doesn't depend on the choice of the Σ-manifold Q bounding the obstruction P' and is determined only by the bordism class $[P]$. So we come to the conclusion:

Lemma 2.3.3 *If the element $k(P) \in MG_*^\Sigma \otimes \mathbf{Z}/2$ is not trivial then the admissible product structure $\mu(Q)$ is not commutative for every choice of the Σ-manifold Q.* □

Now we consider the simple example when it is so. We choose the singularity $\Sigma_0 = (P)$, $[P] = 2$, in the bordism theory $MU_*(\cdot)$. Then we obtain $[P']_{\Sigma_0} = \theta_1 = 0$ as above. Let us take the disk D^2, bounding the circle S^1, $[S^1] = \theta_1$. It is clear that there exists a Σ_0-manifold Q bounding the manifold P' such that the corresponding Σ_j-manifold B is bordant to the manifold $\mathbf{CP}^1$. So the formula (2.9) gives the equality

$$\mu(x \otimes y) - (-1)^{\dim x \dim y} \chi_* \mu(y \otimes x) = \mu(\mu(\beta(x) \otimes \beta(y)) \otimes \left[\mathbf{CP}^1\right]).$$

Note 2.3.1 *The obstructions $[P']_\Sigma$, $[B]_\Sigma$ may be computed in terms of the Steenrod-tom Dick operations, as was proved by O.K.Mironov* [67], [68].

2.4 Associativity for the case of one singularity

Let us determine the Σ-manifold whose cobordism class is the obstruction to associativity of the admissible product structure $\mu(Q)$ in the bordism theory $MG_*^\Sigma(\cdot)$ (when the product structure is commutative)

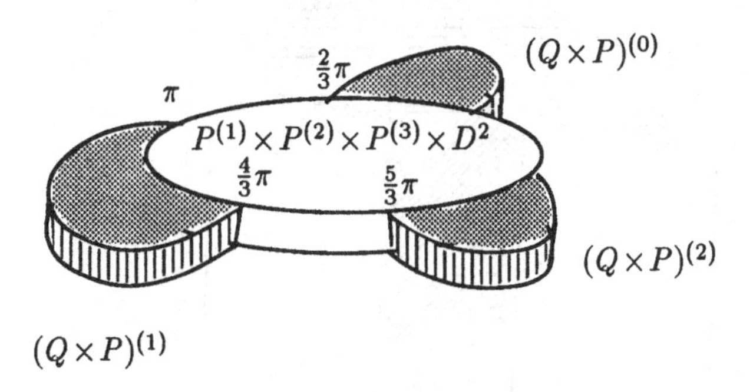

Figure 2.5: Σ-manifold Δ.

as a Σ-manifold by setting

$$\delta\Delta = \delta^{(0)}\Delta \cup \delta^{(1)}\Delta \cup \delta^{(2)}\Delta,$$

$$\delta^{(i)}\Delta = P^{(1)} \times P^{(2)} \times P^{(3)} \times \left[\frac{(2i+1)\pi}{3}, \frac{(2i+2)\pi}{3}\right], \quad i = 0,1,2,$$

$$\beta\Delta = (-1)^p P^{(1)} \times P^{(2)} \left[0, \frac{\pi}{3}\right] \cup P^{(1)} \times P^{(3)} \times \left[\frac{2\pi}{3}, \pi\right]$$
$$\cup (-1)^p P^{(2)} \times P^{(3)} \times \left[\frac{4\pi}{3}, \frac{5\pi}{3}\right].$$

Then we glue the manifold

$$(-1)^{p(i+1)+1} (Q \times P)^{(i)}$$

to the Σ-manifold Δ along every boundary component $\delta^{(i)}\Delta$, for i = 0,1,2, such that the manifold obtained Δ_1 turns out to be a Σ-manifold, that is (see Figure 2.5),

$$\delta^{(0)}\Delta = (-1)^p \left(P^{(1)} \times P^{(2)} \times \left[\frac{\pi}{3}, \frac{2\pi}{3}\right]\right) \times P^{(3)} = (-1)^{p+1}\delta\,(Q \times P)^{(0)},$$

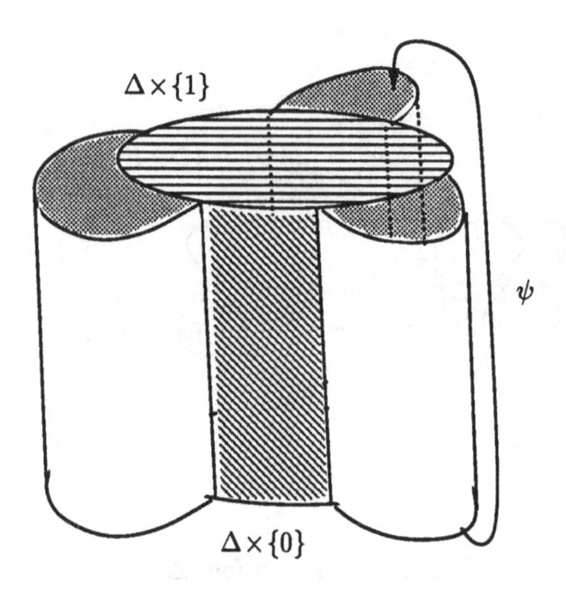

Figure 2.6: Σ-manifold Γ.

$$\delta^{(1)}\Delta = P^{(1)} \times P^{(3)} \times \left[\pi, \frac{4\pi}{3}\right] \times P^{(2)} = (-1)^{p+1}\delta\left(Q \times P\right)^{(1)},$$

$$\delta^{(2)}\Delta = (-1)^{p}\left(P^{(2)} \times P^{(3)} \times \left[\frac{5\pi}{3}, 2\pi\right]\right) \times P^{(1)} = (-1)^{p+1}\delta\left(Q \times P\right)^{(2)}.$$

The diffeomorphism $\psi : \Delta_1 \longrightarrow \Delta_1$ is defined in the following way. It is determined on the part Δ of the manifold Δ_1 by the formula

$$\psi(p_1, p_2, p_3, r, \phi) = (p_3, p_1, p_2, r, \phi + \frac{2}{3}\pi),$$

where (r, ϕ) are the polar coordinates on the disk D^2. Then we put

$$\psi(-1)^{p(i+1)+1}\left(Q \times P\right)^{(i)} = (-1)^{p(i+2)+1}\left(Q \times P\right)^{(i+1)}.$$

We are able now to construct the Σ-manifold Γ as follows (see Figure 2.6):

$$\Gamma = \Delta_1 \times I / \left(\psi(\Delta_1 \times \{0\}) = \Delta_1 \times \{1\}\right).$$

Note 2.4.1 *Let $\Sigma_i = (P_1, \ldots, P_i)$ for $i = 1, \ldots, k$; and let the manifolds Q_i be Σ_i-manifolds such that $\delta Q_i = P_i'$. We note that the above construction determines the Σ_i-manifolds $\Gamma_i = \Gamma(P_i)$ for every $i = 1, \ldots, k$.* □

Theorem 2.4.1 *Let $[P']_\Sigma = 0$ and Q be a Σ-manifold such that $\delta Q = P'$ determining the admissible product structure $\mu(Q)$ in the bordism theory $MG_*^\Sigma(\cdot)$. Then if $[B]_\Sigma = 0, [\Gamma]_\Sigma = 0$, then the product structure $\mu(Q)$ is commutative and associative.*

Proof. The product μ is commutative by Theorem 2.3.1. We consider some Σ-manifolds M^m, N^n, L^l. To prove associativity of the product we use commutativity and compare the following Σ-manifolds:

$$(M \cdot N) \cdot L, \quad (-1)^{(n+l)m}(L \cdot N) \cdot M.$$

We are going to construct a corresponding *bordism* between them in three steps.

1. Here we construct several manifolds required. Let's take the cylinder $D^2 \times I$ with the coordinates (r, ϕ, t), where (r, ϕ) are the polar ones on the disk D^2. Then we consider two curves γ_1, γ_2 on the side of the cylinder $S^1 \times I$ which are determined by the equations

$$\gamma_1 : \frac{\pi}{3}t + \phi - \frac{\pi}{3} = 0, \quad \gamma_2 : \frac{5\pi}{3}t + \phi - \frac{2\pi}{3} = 0$$

(where $0 \leq t \leq 1$, $0 \leq \phi \leq 2\pi$), and the family of horizontal intervals between the curves γ_1 and γ_2:

$$I_t = \left\{ \frac{\pi}{3} - \frac{\pi}{3}t \leq \phi \leq \frac{2\pi}{3} + \frac{2\pi}{3}t \mid 0 \leq t \leq 1 \right\}.$$

Then we choose the family of curves J_τ, depending smoothly on τ on the disk $D^2 \times \{1\}$, such that

i) $J_0 = J_1$,

ii) $J_1 = \left\{ r = 1, \pi \leq \phi \leq \frac{4\pi}{3}, t = 1 \right\}$,

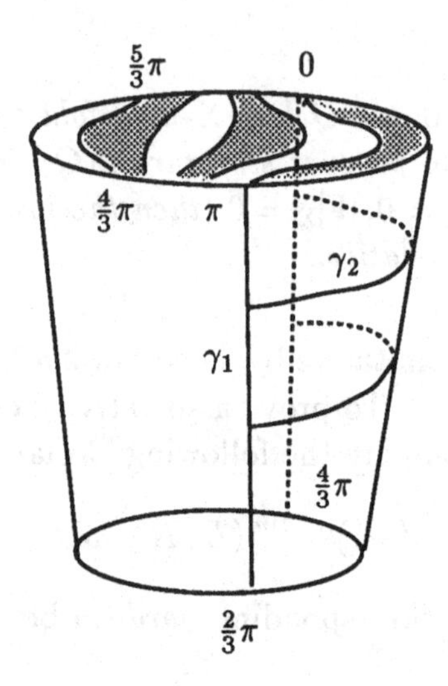

Figure 2.7: $D^2 \times I$.

iii) $J_\tau \neq J_{\tau'}$ then $\tau \neq \tau'$,

iv) $\bigcup_{0 \leq \tau \leq 1} J_\tau = D^2$,

v) $J_\tau(0) = \frac{5\pi}{3} + (1-\tau)\frac{1}{3}\pi, \quad J_\tau(1) = \frac{\pi}{3} + (1-\tau)\frac{1}{3}\pi.$

See Figure 2.7. Now we glue the Σ-manifold

$$(-1)^p Q \times P \times [0,2]$$

to the cylinder

$$P^{(1)} \times P^{(2)} \times P^{(3)} \times D^2 \times I,$$

by identifying the following manifolds:

$$(-1)^p Q \times P \times \{\xi\} = \left\{ \begin{array}{c} (P^{(3)} \times P^{(2)} \times I_\xi) \times P^{(1)} \\ \text{for } 0 \leq \xi \leq 1, \\ \\ (P^{(3)} \times P^{(2)} \times J_{2-\xi}) \times P^{(1)} \\ \text{for } 1 \leq \xi \leq 2. \end{array} \right\} \quad (2.12)$$

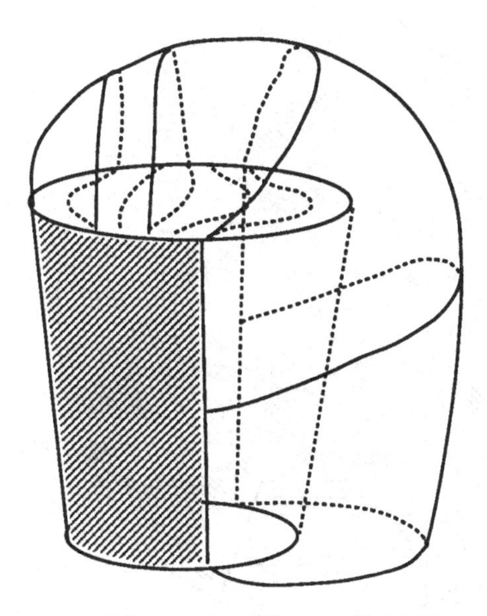

Figure 2.8: Σ-manifold Π_0.

The manifold obtained is denoted by Π_0 (see Figure 2.8)

Finally we glue the manifold

$$Q \times P \times [0,1] \cup Q \times \beta Q \cup (-1)^p Q \times P \times [2,3]$$

to the Σ-manifold Π_0, where

$$\delta\beta Q = P^{(1)} \cup (-1)^p P^{(2)}, \quad Q \times P \times \{1\} = Q \times P^{(1)},$$
$$Q \times P \times \{2\} = Q \times P^{(2)},$$

as follows:

$$\delta Q \times \beta Q = (-1)^p \beta(Q \times P) \times [1,2],$$

$$\delta Q \times P \times \{\eta\} = \left\{ \begin{array}{l} (P^{(1)} \times P^{(3)} \times \left[\pi, \frac{4\pi}{3}\right] \times P^{(2)} \times \{\eta\}, \\ \qquad \text{for } 0 \leq \eta \leq 2, \\ \\ (P^{(2)} \times P^{(3)} \times \left[\frac{5\pi}{3}, 2\pi\right] \times P^{(1)} \times \{3-\eta\}, \\ \qquad \text{for } 2 \leq \eta \leq 3. \end{array} \right\} \quad (2.13)$$

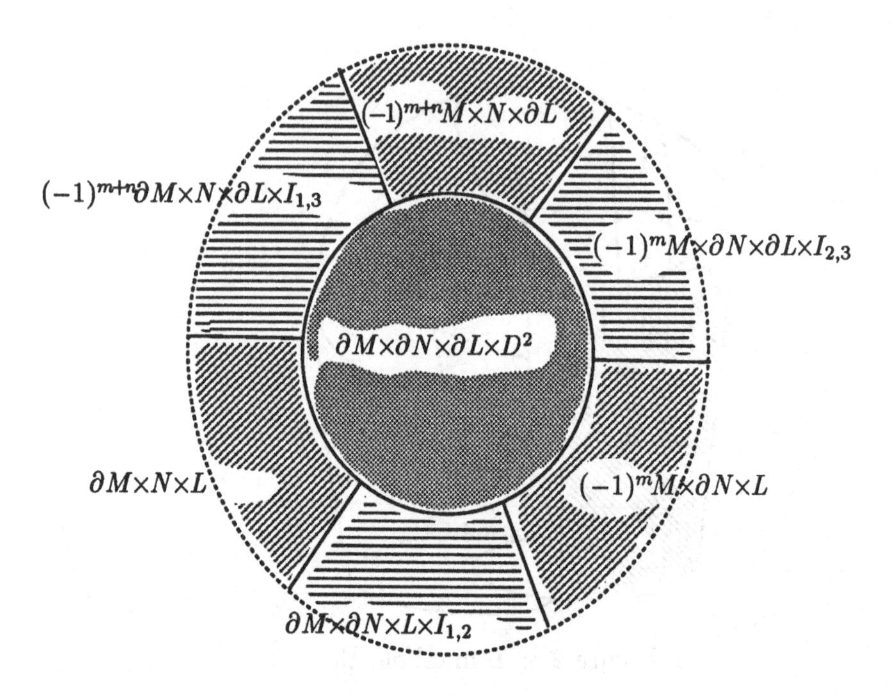

Figure 2.9: A boundary of $M \times N \times L$.

The manifold obtained is denoted by Π (see Figure 2.10(i)).

2. We suppose that the Σ-manifolds M, N, L have collars along their boundaries. (We consider them as ordinary manifolds with nonempty boundaries.) So the boundary of the manifold $M \times N \times L$ may be decomposed as follows:

$$\begin{gathered}
\partial(M \times N \times L) = \partial M \times N \times L \cup \partial M \times \partial N \times I_{1,2} \times L \\
\cup\, (-1)^m M \times \partial N \times L \cup\, (-1)^m M \times \partial N \times \partial L \times I_{2,3} \\
\cup\, (-1)^{m+n} M \times N \times \partial L \cup (-1)^{m+n} \partial M \times N \times \partial L \times I_{1,3} \\
\cup\, \partial M \times \partial N \times \partial L \times D^2.
\end{gathered}$$

A gluing scheme is shown in Figure 2.9.

We glue the manifolds

$$\beta M \times \beta N \times L \times Q, \quad M \times \beta N \times \beta L \times Q, \quad \beta M \times N \times \beta L \times Q$$

to the product $M \times N \times L$ by identifying

$$\beta M \times \beta N \times L \times \delta Q, \quad M \times \beta N \times \beta L \times \delta Q, \quad \beta M \times N \times \beta L \times \delta Q$$

with

$$\partial M \times \partial N \times L \times I_{1,2}, \quad M \times \partial N \times \partial L \times I_{2,3}, \quad \partial M \times \partial N \times L \times I_{1,3}$$

respectively. Here we suppose that

$$\delta Q = P^{(1)} \times P^{(2)} \times I.$$

Note that the manifold obtained U is a Σ-manifold with boundary

$$\beta M \times \beta N \times \beta L \times \Delta.$$

The manifold $(M \cdot N) \cdot L$ is obtained by the gluing of the manifolds U with the product

$$\beta M \times \beta N \times \beta L \times \Pi$$

along the manifold $\beta M \times \beta N \times \beta L \times \Delta$; see Figure 2.10.

A similar procedure gives the manifold

$$(-1)^{m(n+l)}(L \cdot N) \cdot M.$$

That is, we are to use a new manifold Π_1 (instead of the manifold Π), which we construct as follows. We define the diffeomorphism

$$T : P^{(1)} \times P^{(2)} \times P^{(3)} \times D^2 \times I \longrightarrow P^{(1)} \times P^{(2)} \times P^{(3)} \times D^2 \times I$$

by the equality

$$T(p_1, p_2, p_3, r, \phi, t) = (\Psi(p_1, p_2, p_3, r, \phi), t).$$

The next gluing of the manifolds

$$(-1)^p Q \times P \times [2,3], \quad Q \times P \times [0,1] \cup Q \times \beta Q \cup Q \times P \times [2,3]$$

must be twisted by the diffeomorphism T.

i)

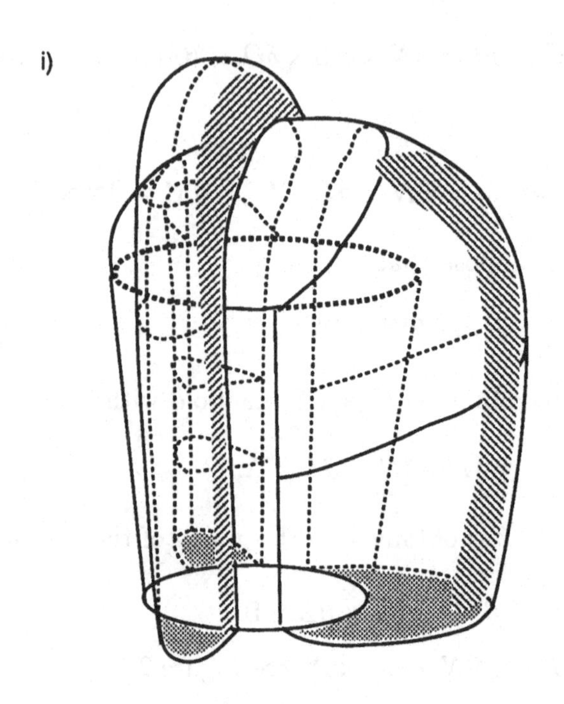

ii)

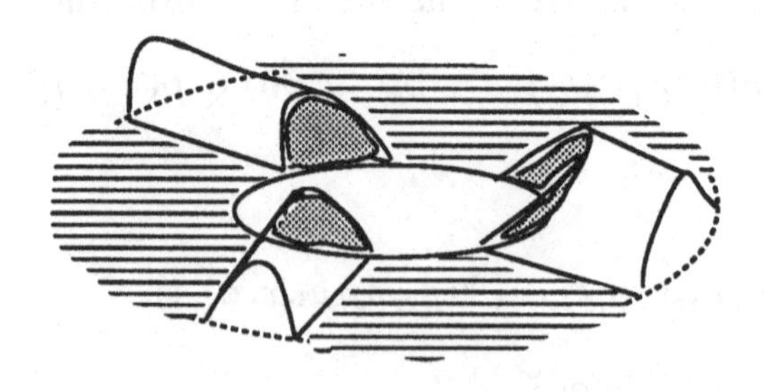

Figure 2.10: A gluing of the Σ-manifold $(M \cdot N) \cdot L$.

As a result we obtain the diffeomorphism $T_1 : \Pi_1 \longrightarrow \Pi_1$ which coincides with the diffeomorphism $\Psi : \Delta \longrightarrow \Delta$ under restriction to Δ.

3. We consider two cylinders

$$(M \cdot N) \cdot L \times I_1, \quad (-1)^{m(n+l)}(L \cdot N) \cdot M \times I_2.$$

Let us glue them by the cylinder $U \times I_3$:

$$U \times \{1\} = U \times \{0\} \subset (M \cdot N) \cdot L \times \{0\},$$

$$U \times \{0\} = U \times \{0\} = (-1)^{m(n+l)}(L \cdot N) \cdot M \times \{0\}.$$

Then the manifolds

$$\beta M \times \beta N \times \beta L \times \Pi \times \{0\} \subset \beta\left((M \cdot N) \cdot L \times \{0\}\right),$$

$$\beta M \times \beta N \times \beta L \times \Pi_1 \times \{0\} \subset \beta\left((-1)^{m(n+l)}(N \cdot L) \cdot M \times \{0\}\right)$$

should be glued by the diffeomorphism which is determined by the formula

$$(x, y, z, \xi) \longrightarrow (x, y, z, T_1(\xi)).$$

The manifold obtained $A(M, N, L)$ is a Σ-manifold with the boundary

$$(M \cdot N) \cdot L \cup \beta M \times \beta N \times \beta L \times \Gamma \cup (-1)^{m(n+l)}(N \cdot L) \cdot M.$$

So if $[\Gamma]_\Sigma = 0$, then the 3-linear construction $\mathfrak{A}(M, N, L)$ is determined, its boundary equal to the manifold

$$\delta\mathfrak{A}(M, N, L) = (M \cdot N) \cdot L \cup (-1)^{m(n+l)}(N \cdot L) \cdot M.$$

To finish the proof we have to glue to $\mathfrak{A}(M, N, L)$ the bilinear construction $\mathfrak{B}(N \cdot L, M)$. Lemma 2.1.4 gives us the associativity of the product $\mu(Q)$. $\square$

Further we will use the following simple property of the manifolds Γ.

Lemma 2.4.2 *The following equality is true in the group MG^{Σ}_{*}:*

$$3[\Gamma]_{\Sigma} = 0. \quad \square$$

Note 2.4.2 *The admissible product structure may be associative without a commutativity condition. In particular it can be so when the obstruction manifolds P', B, and Γ are bordant to manifolds without singularities. For example, the noncommutative product in the bordism theory $MU^{\Sigma_0}_{*}(\cdot)$ where $\Sigma_0 = (P)$, $[P] = 2$, is associative; see* [67, 68]. $\square$

2.5 General case

Let $\Sigma_i = (P_1, \ldots, P_i)$ for $i = 1, \ldots, k$. We suppose as before that $[P_i']_{\Sigma_i} = 0$. We take the Σ_i-manifolds Q_i such that $\delta Q_i = P_i'$ which determine the admissible external product structure $\mu_i = \mu_i(Q_1, \ldots, Q_i)$ in the theory $MG^{\Sigma_i}_{*}(\cdot)$ for every $i = 1, \ldots, k$. Note that we have already determined the following Σ_i-manifolds:

$$B = B(P_i), \quad \Gamma = \Gamma(P_i).$$

Theorem 2.5.1 *Let $\mu_k = \mu_k(Q_1, \ldots, Q_k)$ be an admissible product structure in the bordism theory $MG^{\Sigma_k}_{*}(\cdot)$ If the obstructions*

$$[B_i]_{\Sigma_i} = 0, \quad [\Gamma_i]_{\Sigma_i} = 0$$

are trivial for all $i = 1, \ldots, k$ then the product structure μ_k in the bordism theory $MG^{\Sigma_k}_{*}(\cdot)$ is commutative and associative.

Proof. We begin by clarifying the scheme of the proof. We have the bilinear construction $\mathfrak{B}_i$ and 3-linear construction $\mathfrak{A}_i$ on the category $\mathfrak{M}_i$ for every $i = 1, \ldots, k$, which have the following properties:

$$1) \qquad \delta\mathfrak{B}_i(M, N) = (\mathrm{m}_i(M, N) \cup (-1)^{mn}\mathrm{m}_i(N, M))$$
$$\cup(\mathfrak{B}_i(\delta M, N) \cup \mathfrak{B}_i(M, \delta N));$$

2) $$\delta\mathfrak{A}_i(M,N,L) = \mathfrak{m}_i(\mathfrak{m}_i(M,N),L) \cup \mathfrak{m}_i(M,\mathfrak{m}_i(N,L))$$
$$\cup\, \mathfrak{A}_i(\delta M,N,L) \cup \mathfrak{A}_i(M,\delta N,L) \cup \mathfrak{A}_i(M,N,\delta L),$$

where the constructions

$$\mathfrak{A}_i(\delta M,N,L) \quad \textit{and} \quad \mathfrak{A}_i(M,\delta N,L)$$

are glued along the construction $\mathfrak{A}_i(\delta M,\delta N,L)$, *the constructions*

$$\mathfrak{A}_i(M,\delta N,L), \quad \mathfrak{A}_i(M,N,\delta L)$$

along $\mathfrak{A}_i(M,\delta N,\delta L)$, *and the constructions*

$$\mathfrak{A}_i(\delta M,N,L), \quad \mathfrak{A}_i(M,N,\delta L)$$

along $\mathfrak{A}_i(\delta M,N,\delta L)$;

3) *If* M,N,L *are* Σ_j*-manifolds, where* $j<i$, *then*

$$\mathfrak{A}_i(M,N,L) = \mathfrak{A}_j(M,N,L), \quad \mathfrak{B}_i(M,N) = \mathfrak{B}_j(M,N);$$

4) *If one of the manifolds* M,N *has no singularities then*

$$\mathfrak{B}_i(M,N) = M \times N \times I;$$

5) *If one of the manifold* M, N, L *(*N *for example) has no singularities then*

$$\mathfrak{A}_i(M,N,L) = (-1)^{nl}\mathfrak{m}_i(M,L) \times N \times I;$$

6) *The constructions* $\mathfrak{B}_0$ *and* $\mathfrak{A}_0$ *on the category* $\mathfrak{M}$ *are defined as*

$$\mathfrak{B}_0(M,N) = M \times N \times I, \quad \mathfrak{A}_0(M,N,L) = M \times N \times L \times I.$$

Let us describe an induction step for defining the construction $\mathfrak{B}_i$. Let M^m and N^n be Σ_i-manifolds and put that $\alpha(n) = n(p_i+1)+1$, where $p_i = \dim P_i$. For simplicity we suppose that $\delta M = \emptyset$, $\delta N = \emptyset$. Let's apply the construction $\mathfrak{B}_{i-1}$ at the Σ_i-manifolds X,Y; the latter may be considered as the Σ_{i-1}-manifolds $\widehat{X},\widehat{Y}$ as follows:

$$\partial_j\widehat{X} = \partial_j X, \quad \partial_j\widehat{Y} = \partial_j Y, \quad j=1,\ldots,i-1,$$

$$\delta\widehat{X} = \delta X \cup \partial_i X, \quad \delta\widehat{Y} = \delta Y \cup \partial_i Y.$$

Note that all our manifolds have collars. So we have

$$\mathfrak{m}_i(M,N) = \mathfrak{m}_{i-1}(M,N) \cup (-1)^{\alpha(n)}\mathfrak{m}_{i-1}(\mathfrak{m}_{i-1}(\beta_i M, \beta_i N), Q_i).$$

Consider two cylinders

$$C_1 = \mathfrak{m}_i(M,N) \times I_1, \quad C_2 = (-1)^{mn}\mathfrak{m}_i(N,M) \times I_2.$$

We glue the cylinders C_1 and C_2 with $\mathfrak{m}_{i-1}(M,N)$ by identifying the manifolds

$$\mathfrak{m}_{i-1}(M,N) \times \{0\} = \mathfrak{m}_{i-1}(M,N) \hookrightarrow \delta\mathfrak{B}_{i-1}(M,N),$$

$$(-1)^{mn}\mathfrak{m}_{i-1}(M,N) \times \{1\} = (-1)^{mn}\mathfrak{m}_{i-1}(M,N) \hookrightarrow \delta\mathfrak{B}_{i-1}(N,M).$$

The construction obtained is denoted by $\mathfrak{B}_i^{(1)}(M,N)$. Note that the boundary of the construction $\mathfrak{B}_{i-1}(M,N)$ is equal to

$$\mathfrak{B}_{i-1}(M,N) = \mathfrak{m}_{i-1}(M,N) \cup (-1)^{mn}\mathfrak{m}_{i-1}(N,M)$$

$$\cup \mathfrak{B}_{i-1}(\partial_i M, N) \cup \mathfrak{B}_{i-1}(\partial_i M, \partial_i N) \times I \cup \mathfrak{B}_{i-1}(M, \partial_i N)$$

$$= \mathfrak{m}_{i-1}(M,N) \cup (-1)^{mn}\mathfrak{m}_{i-1}(N,M) \cup \mathfrak{B}_{i-1}(\beta_i M, N) \times P_i^{(1)}$$

$$\cup \mathfrak{B}_{i-1}(\beta_i M, \beta_i N) \times P_i' \cup \mathfrak{B}_{i-1}(M, \beta_i N) \times P_i^{(2)}.$$

Then we glue the above construction to $\mathfrak{B}_i^{(1)}(M,N)$:

$$\mathfrak{m}_{i-1}(\mathfrak{B}_{i-1}(\beta_i M, \beta_i N), \delta Q_i) = \mathfrak{B}_{i-1}(\beta_i M, \beta_i N) \times P_i' \subset \delta\mathfrak{B}_i^{(1)}(M,N);$$

$$(-1)^{\alpha(n)}\mathfrak{m}_{i-1}(\mathfrak{B}_{i-1}(\beta_i M, \beta_i N), Q_i)$$

$$= (-1)^{\alpha(n)}\mathfrak{m}_{i-1}(\mathfrak{B}_{i-1}(\beta_i M, \beta_i N), Q_i) \times \{0\} \subset \mathfrak{m}_i(M,N) \times \{0\} \subset C_1.$$

The resulting bilinear construction is denoted by $\mathfrak{B}_i^{(2)}(M,N)$; see Figure 2.11. The manifold $\mathfrak{B}_i^{(2)}(M,N)$ may be considered as a Σ_i-manifold with the boundary

$$\delta\mathfrak{B}_i^{(2)}(M,N) = \mathfrak{m}_{i-1}(M,N) \cup (-1)^{mn}\mathfrak{m}_{i-1}(N,M) \cup \mathfrak{m}_i(\mathfrak{m}_i(\beta_i M, \beta_i N), B_i),$$

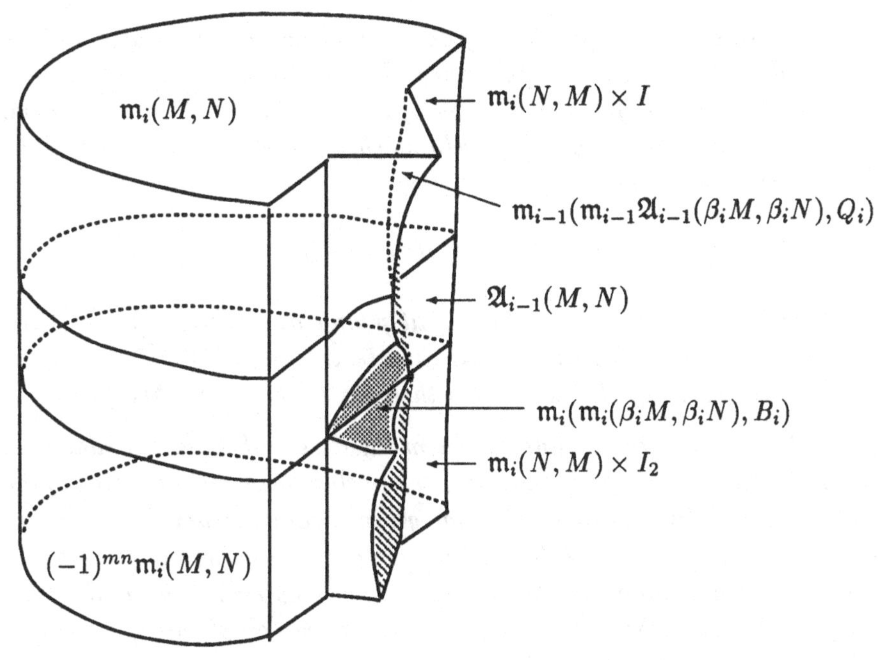

Figure 2.11: Construction $\mathfrak{B}_i(M,N)$.

where $B_i = B(P_i)$ is the corresponding obstruction to commutativity. We take the Σ_i-manifold D_i which bounded the obstruction B_i and then glue the construction

$$\mathfrak{m}_i(\mathfrak{m}_i(\beta_i M,\beta_i N),D_i)$$

to $\mathfrak{B}_i^{(2)}(M,N)$. The result is the bilinear construction $\mathfrak{B}_i(M,N)$, possessing all the above properties. The construction $\mathfrak{A}_i(M,N,L)$ may be determined in a similar way using the proof of Theorem 2.4.1. □

Note 2.5.1 1) *Theorem 2.5.1 provides the sufficient conditions for commutativity and associativity of the admissible product structure. In several cases it is sufficient to verify that the groups*

$$MG^{\Sigma_k}_{2p_k+2},\quad MG^{\Sigma_k}_{3p_k+3}$$

haven't torsion of order two and three respectively.

2) *There exists the problem of describing the operation algebra of a (co)-bordism theory with singularities. The solution seems to be not so simple. Only a handful of examples of solving the problem are known. The results were obtained by purely algebraic methods. In particular, the operation algebras*

$$k(n)^*(k(n)), \quad P\langle n\rangle^*(P\langle n\rangle)$$

have been computed completely for all p and n; see [51], [120], [121]. *It would be very interesting to investigate the connection between the operation algebras of the cobordism theories $MG^*(\cdot)$ and $MG^*_\Sigma(\cdot)$.*

3) *It is natural to consider bordism theories with deep singularities, i.e. where the manifold-singularities themselves have singularities of the previous level. (K.Nowinsky* [78] *has given a rather general description of such theories.) It is not difficult to give corresponding definitions and examples; one of them will be considered in Chapter 4. The multiplicativity problem in these theories is much more complicated, moreover there are some obstructions to equipping such a theory with deep singularities with module structure over the original bordism $MG^\Sigma_*(\cdot)$ theory with the usual singularities. The above obstructions are certainly trivial if a ring structure of the spectrum MG^Σ can be extended up to a H_∞-structure. To investigate the problem seems to be interesting.*

Note also that all the bordism theories $MG^\Sigma_*(\cdot)$ from the examples 1–4 (section 1.2) do have commutative and associative admissible product structures. The same is true for the theory $P\langle 0\rangle_*(\cdot)$ for every prime p and for the theories $P\langle n\rangle_*(\cdot)$ for $p > 2, n \geq 1$. Nevertheless the theory $P\langle n\rangle_*(\cdot)$ for $p = 2$, $n \geq 1$ doesn't admit the commutative product structure, there exists only associative product structure; see [67], [68].

2.6 Product structure in the Σ-SSS

The above geometric construction of a product structure in a bordism theory with singularities may be extended up to a product structure in the corresponding Σ-singularities spectral sequence. Actually we can do without the direct construction because the product structure in the spectral sequence appears after identifying it with the Adams-Novikov spectral sequence. This product structure is also generated by an admissible one in the bordism theory with singularities.

Our plan is the following. First we intend to formulate the general problem of constructing the product structure in the Σ-*SSS* and then we are going to explicitly consider the case of one singularity.

The $\Sigma\Gamma(k)$-manifolds and the bordism theory of these manifolds as well as the transformations

$$MG_*(\cdot) \xleftarrow{\gamma(1)} MG_*^{\Sigma\Gamma(1)}(\cdot) \longleftarrow \cdots \longleftarrow MG_*^{\Sigma\Gamma(k-1)}(\cdot) \xleftarrow{\gamma(k)} \cdots$$

were determined in section 1.4.

The transformations $\gamma(k)$ were determined at the level of the manifolds, in particular every $\Sigma\Gamma(k)$-manifold M has the structure of a $\Sigma\Gamma(n)$-manifold for every $n < k$.

Two $\Sigma\Gamma(n)$- and $\Sigma\Gamma(k)$-structures on the given manifold M (for $n < k$) are *compatible* if the $\Sigma\Gamma(n)$-structure induced by the $\Sigma\Gamma(k)$-structure coincides with the first one.

Let $p < q$ be integers.

Definition 2.6.1 *The manifold M is called a $\Sigma\Gamma(p,q)$-manifold if*

(i) *it has the $\Sigma\Gamma(p)$-structure,*

(ii) *its boundary ∂M has the partition*

$$\partial M = \partial^{(0)} M \cup \partial^{(1)} M,$$

such that $\partial(\partial^{(0)} M) = -\partial(\partial^{(1)} M)$, also the manifold $\partial^{(1)} M$ has a $\Sigma\Gamma(q)$-structure which is compatible with the original $\Sigma\Gamma(p)$-structure. □

The manifold $\partial^{(0)} M$ possesses the induced $\Sigma\Gamma(p,q)$-structure by definition and is called the *boundary of the $\Sigma\Gamma(p,q)$-manifold M*. Note that if $p > p'$, $q > q'$, then the $\Sigma\Gamma(p,q)$-manifold M has the structure of a $\Sigma\Gamma(p',q')$-manifold. The notion of bordism of such manifolds may be determined in a standard way. Thus we have the corresponding *bordism theory $MG_*^{\Sigma\Gamma(p,q)}(\cdot)$*.

Note 2.6.1 *These bordism theories are natural generalizations of the bordism theories with ordinary singularities. In particular, it is clear that $MG_*^{\Sigma\Gamma(0,1)}(\cdot) = MG_*^{\Sigma}(\cdot)$, and the theory $MG_*^{\Sigma\Gamma(0,p)}(\cdot)$ coincides with the bordism theory $MG_*^{\Sigma^p}(\cdot)$ (where Σ^p is the sequence of the p-fold products of the manifolds $P_i, i = 1, 2, \dots$). The theory $MG_*^{\Sigma\Gamma(p,q)}(\cdot)$ may be considered as the bordism theory of $\Sigma\Gamma(p)$-manifolds with Σ^q-singularities.* □

The following transformation is defined for all numbers $p > p'$, $q > q'$:

$$\eta : MG_*^{\Sigma\Gamma(p,q)}(\cdot) \longrightarrow MG_*^{\Sigma\Gamma(p',q')}(\cdot). \tag{2.14}$$

It forgets $\Sigma\Gamma(p,q)$-structure up to $\Sigma\Gamma(p',q')$-structure. By taking the singularity manifold $\partial^{(1)}M$ of the given $\Sigma\Gamma(p,q)$-manifold M we get the transformation

$$d : MG_*^{\Sigma\Gamma(p,q)}(\cdot) \longrightarrow MG_*^{\Sigma\Gamma(q)}(\cdot).$$

To formulate the multiplicativity problems in the Σ-*SSS* we are going to use the terminology of *spectral systems* (see [38]). For this we determine the bordism theory pairings

$$\phi_r : MG_*^{\Sigma\Gamma(p,p+r)}(\cdot) \otimes MG_*^{\Sigma\Gamma(q,q+r)}(\cdot) \longrightarrow MG_*^{\Sigma\Gamma(p+q,p+q+r)}(\cdot)$$

for all p, q, r, which have to be compatible with the transformations η and d. The general case of many singularities leads us to a very complicated construction. Here we prefer to restrict attention to the case when $\Sigma = (P)$; we shall further assume either that the number $k = \dim P$ is even, or that $2[P] = 0$ in the group MG_*. We have the isomorphism

$$MG_n^{\Sigma\Gamma(p,p+r)}(X,Y) \cong MG_{n-kp}^{\Sigma^r}(X,Y),$$

where $\Sigma^r = (P^r)$. Here it is convenient to decompose a $\Sigma\Gamma(p,p+r)$-manifold into the direct product $M \times P^p$, where M is Σ^r-manifold.

If $p \geq p'$, $r \geq r'$, then the transformation

$$\eta : MG_*^{\Sigma\Gamma(p,p+r)}(\cdot) \longrightarrow MG_*^{\Sigma\Gamma(p',p'+r')}(\cdot)$$

is induced by the map which translates the direct product $M \times P^p$ (where M is a Σ^r-manifold) to the direct product $M \times P^{p'}$ (where the manifold M should be considered as a $\Sigma^{r'}$-manifold).

The transformation

$$\delta : MG_*^{\Sigma\Gamma(p,p+r)}(\cdot) \longrightarrow MG_*^{\Sigma\Gamma(p+r,p+r+r')}(\cdot)$$

is induced by the map which translates the direct product $M \times P^p$ (where M is a Σ^r-manifold) to the direct product

$$\beta^{(r)}M \times P^{p+r},$$

where $\beta^{(r)}M$ is considered as a $\Sigma^{r'}$-manifold.

Now we are going to construct the pairings

$$\phi_r : MG_{m-kp}^{\Sigma^r}(X) \otimes MG_{n-kq}^{\Sigma^r}(X_1) \longrightarrow MG_{n+m-k(p+q)}^{\Sigma^r}(X \wedge X_1),$$

which can be determined when the bordism theories $MG_*^{\Sigma^r}(\cdot)$ possess admissible product structures.

It is the bordism class of the Σ^r-manifold

$$(P^r)' = (P^r)^{(1)} \times (P^r)^{(2)} \times I$$

that is the obstruction to the existence of the admissible product structure in the theory $MG_*^{\Sigma^r}(\cdot)$.

Lemma 2.6.2 *If $[P']_\Sigma = 0$ in the group MG_*^Σ then the element $[(P^r)']_\Sigma$ is also zero in the group MG_*^Σ for every $r = 1, 2, \ldots$.*

Proof. According to the above definitions we have

$$(P^r)' = P^{(1)} \times \ldots \times P^{(r)} \times P^{(r+1)} \times \ldots \times P^{(2r)} \times I.$$

Let us choose the partition of the interval $I = [0,1]$ into r parts,

$$I = \bigcup_{i=1}^{r} \left[\frac{i-1}{r}, \frac{i}{r}\right].$$

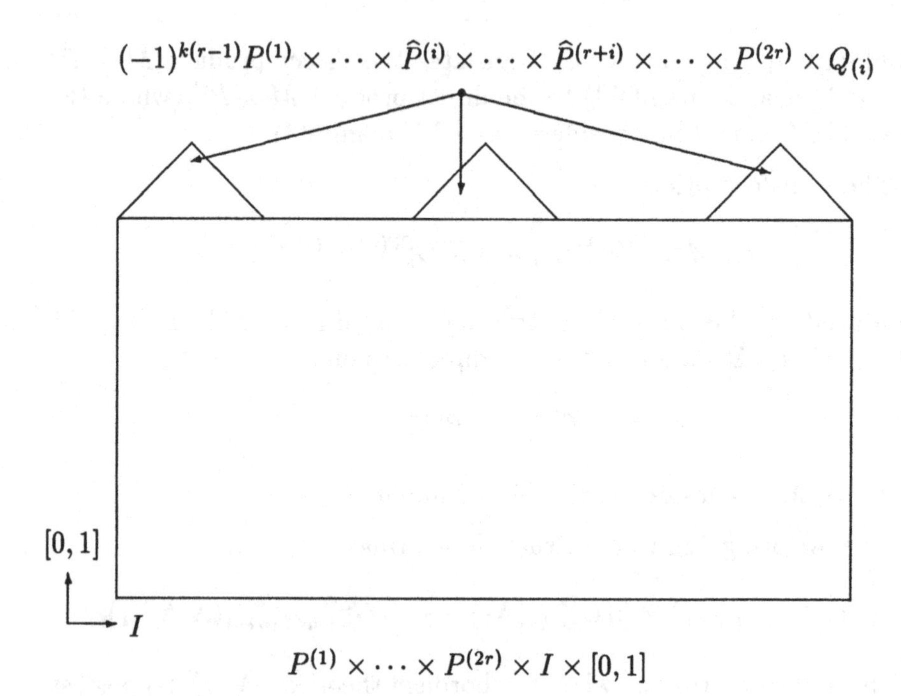

Figure 2.12: $Q^{(r)}$.

Now let's take some Σ-manifold Q , such that

$$\delta Q = P' = P^{(1)} \times P^{(2)} \times I.$$

Then we take r copies of this manifold denoting them by $Q_{(1)}, \ldots, Q_{(r)}$ respectively. We suppose that

$$\delta Q_{(i)} = P^{(i)} \times P^{(r+i)} \times \left[\frac{i-1}{r}, \frac{i}{r}\right].$$

Then the Σ^r-manifold

$$Q^{(r)} = \bigcup_{i=1}^{r} (-1)^{k(r-1)} P^{(1)} \times \ldots \times \widehat{P}^{(i)} \times \ldots \times \widehat{P}^{(i+r)} \times \ldots \times P^{(2r)} \times Q_{(i)}$$

bounds the obstruction $\delta^{(r)} Q^{(r)} = (P^r)'$. □

Theorem 2.2.2 implies that the bordism theory $MG_*^{\Sigma^r}(\cdot)$ has the admissible product structure ϕ^r. Let us examine the Σ^r-manifold $Q^{(r)}$ more carefully.

Lemma 2.6.3 *If* $[P']_\Sigma = 0$ *in the group* MG_*^Σ *then the element* $\left[\beta^{(r)}Q^{(r)}\right]$ *is also zero in the group* $MG_*^{\Sigma^{r-1}}$ *where* $Q^{(r)}$ *is the above* Σ^r*-manifold.*

Proof. Let's construct the manifold which is diffeomorphic to the manifold $Q^{(r)}$. We consider the partition of the interval $I = [0,1]$ into $2r-1$ parts,

$$I = \bigcup_{j=1}^{2r-1} \left[\frac{j-1}{2r-1}, \frac{j}{2r-1}\right],$$

and take the cylinder

$$P^{(1)} \times \ldots \times P^{(r)} \times P^{(r+1)} \times \ldots \times P^{(2r)} \times I \times [0,1].$$

We glue to its top side the manifolds

$$(-1)^{k(r-1)} P^{(1)} \times \ldots \times \widehat{P}^{(i)} \times \ldots \times \widehat{P}^{(i+r)} \times \ldots \times P^{(2r)} \times Q_{(i)},$$

$$\delta Q_{(i)} = P^{(1)} \times \ldots \times P^{(2r)} \times \left[\frac{2(i-1)}{2r-1}, \frac{2i-1}{2r-1}\right]$$

(where $i = 1, \ldots, r$), by identifying the manifolds

$$(-1)^{k(r-1)} P^{(1)} \times \ldots \times \widehat{P}^{(i)} \times \ldots \times \widehat{P}^{(i+r)} \times \ldots \times P^{(2r)} \times \delta Q_{(i)}$$

$$= P^{(1)} \times \ldots \times P^{(2r)} \times \left[\frac{2(i-1)}{2r-1}, \frac{2i-1}{2r-1}\right] \times \{0\}.$$

We denote the resulting Σ^r-manifold by $\widetilde{Q}^{(r)}$; see Figure 2.12. It is clear that $\widetilde{Q}^{(r)}$ is diffeomorphic to $Q^{(r)}$. Now we consider the Σ^r-manifold $\beta^{(r)}\widetilde{Q}^{(r)}$. According to the above constructions we have

$$\beta^{(r)}\widetilde{Q}^{(r)} = \left(\bigcup_{i=1}^{r-1} (-1)^{k(r-1)} \left(P^{(1)} \times \ldots \times P^{(i-1)} \times \ldots \right.\right.$$

$$\left.\left. \ldots \times P^{(i+r)} \times \ldots \times P^{(2r)} \times \left[\frac{2i-1}{2r-1}, \frac{2i}{2r-1}\right]\right)\right)$$

$$\cup \left(\bigcup_{i=1}^{r} (-1)^{k(r+1)(i-1)} P^{(1)} \times \ldots \times P^{(i-1)} \times \ldots \times P^{(i+r)} \times \ldots \times P^{(2r)} \times \beta Q_{(i)}\right),$$

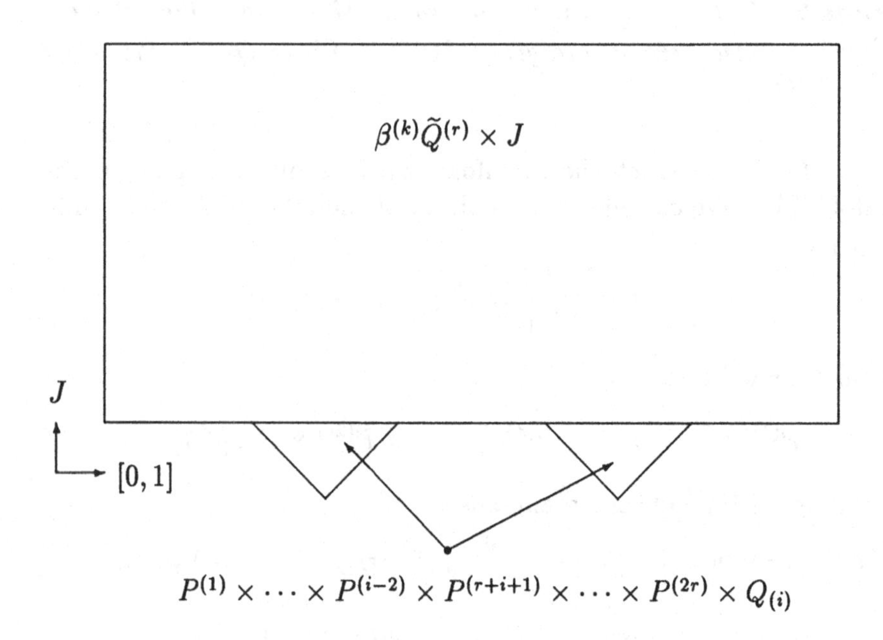

Figure 2.13: $\beta^{(r)}\tilde{Q}^{(r)} \times J$

where $P^{(0)}$ is the point. Let us take the cylinder

$$\beta^{(r)}\tilde{Q}^{(r)} \times J, \quad J = [0,1],$$

and glue the following manifolds to its bottom side:

$$P^{(1)} \times \ldots \times P^{(i-2)} \times P^{(r+i+1)} \times \ldots \times P^{(2r)} \times Q_{(i)},$$

have where $i = 2, \ldots, r$, and $P^{(2r+1)}$ is the point. Then we proceed to the following identification:

$$P^{(1)} \times \ldots \times P^{(i-2)} \times P^{(i+r+1)} \times \ldots P^{(2r)} \times \beta Q_{(i)}$$

$$= P^{(1)} \times \ldots \times P^{(i-1)} \times P^{(i+r)} \times \ldots \times P^{(2r)} \times \left[\frac{2i-3}{2r-1}, \frac{2i-2}{2r-1}\right],$$

using the fact that the direct product

$$P^{(1)} \times \ldots \times P^{(i-1)} \times P^{(i+r)} \times \ldots \times P^{(2r)} \times \left[\frac{2i-3}{2r-1}, \frac{2i-2}{2r-1}\right]$$

is a part of $\beta^{(r)}\tilde{Q}^{(r)}$. The manifold obtained is to be a Σ^{r-1}-manifold with the boundary $\beta^{(r)}\tilde{Q}^{(r)}$; see Figure 2.13.

Let us formulate Douady's conditions for the pairings ϕ_r in terms of the new geometric interpretation (see [38]). These conditions guarantee the existence of the product structure in the Σ-*SSS*.

1. *If* $p \geq p'$, $q \geq q'$, $r \geq r'$, *then the following diagram must be commutative:*

$$\begin{array}{ccc} MG^{\Sigma^r}_{m-kp}(X) \otimes MG^{\Sigma^r}_{n-kq}(Y) & \xrightarrow{\phi_r} & MG^{\Sigma^r}_{m+n-k(p+q)}(X \wedge Y) \\ \downarrow & & \downarrow \\ MG^{\Sigma^{r'}}_{m'-kp'}(X) \otimes MG^{\Sigma^{r'}}_{n'-kq'}(Y) & \xrightarrow{\phi'_r} & MG^{\Sigma^{r'}}_{m'+n'-k(p'+q')}(X \wedge Y) \end{array} \tag{2.15}$$

where $m' = m + k(p' - p)$, $n' = n + k(p' - p)$.

2. *All the homomorphisms of the diagram*

$$\begin{array}{ccc} MG^{\Sigma^r}_{m-kp}(X) \otimes MG^{\Sigma^r}_{n-kq}(Y) & \xrightarrow{\eta\otimes\delta} & MG^{\Sigma^r}_{m-kp}(X) \otimes MG^{\Sigma^1}_{n-kq}(Y) \\ {\scriptstyle\delta\otimes\eta}\downarrow & \searrow^{\phi_r} \quad MG^{\Sigma^r}_{m+n-k(p+q)}(X \wedge Y) \quad \searrow^{\delta} & \downarrow{\scriptstyle\phi_1} \\ MG^{\Sigma^1}_{m-kp}(X) \otimes MG^{\Sigma^r}_{n-kq}(Y) & \xrightarrow{\phi_1} & MG^{\Sigma^1}_{m+n-k(p+q)-1}(X \wedge Y) \end{array} \tag{2.16}$$

satisfy the equality

$$\delta \circ \phi_r = \phi_1 \circ (\delta \otimes \eta) + \phi_1 \circ (\eta \otimes \delta)$$

for all numbers $p, q \geq 0$ *and* $r \geq 1$ *and spaces* X, Y.

Let us verify the *first condition*. Suppose M, N are Σ^r-manifolds. According to the definitions of the transformations η, ϕ_r we have

$$\eta \circ \phi_r\,([M] \otimes [N]) = [M \cdot N \times P^{p+q-p'-q'}],$$

$$\phi_{r'}(\eta \otimes \eta)\,([M] \otimes [N]) = [(M \times P^{p-p'}) \cdot (N \times P^{q-q'})],$$

where the point means the product in the theories $MG^{\Sigma^r}(\cdot)$ and $MG^{\Sigma^{r'}}(\cdot)$. The module structure of these theories over the theory $MG_*(\cdot)$ implies commutativity of (2.15).

Let us verify the *second condition*. According to the definition of the transformations δ, η we get

$$\delta \circ \phi_r ([M] \otimes [N]) = \eta[\beta^{(r)}(M \cdot N)].$$

Note that

$$\left[\beta^{(r)}(M \cdot N\right]_{\Sigma^r} = \left[\beta^{(r)}M \cdot N\right]_{\Sigma^r} + \left[M \cdot \beta^{(r)}N\right]_{\Sigma^r}$$
$$+(-1)^{k(m-k)} \left[\left(\beta^{(r)}M \cdot \beta^{(r)} \cdot N\right) \cdot \widehat{\beta}^{(r)}Q^{(r)}\right]_{\Sigma^r}.$$

Lemma 2.6.3 implies that

$$\left[\widehat{\beta}^{(r)}Q^{(r)}\right]_{\Sigma^r} = 0.$$

So we have the equality

$$\eta\left(\left[\beta^{(r)}(M \cdot N)\right]_{\Sigma^r}\right) = \eta\left(\left[(\beta^{(r)}M) \cdot N\right]_{\Sigma^r}\right) + \eta\left(\left[M \cdot (\beta^{(r)}N)\right]_{\Sigma^r}\right)$$
$$= ((\phi_1 \circ (\delta \otimes \eta) + (\phi_1 \circ (\eta \otimes \delta))([M] \otimes [N]).$$

Thus we have proved the following.

Theorem 2.6.4 *Let the manifold P have even dimension or the element $[P]$ have order two in the group MG_*, and the obstruction $[P']_\Sigma \in MG_*^\Sigma$ to the existence of the admissible product structure in the bordism theory $MG_*^\Sigma(\cdot)$, $\Sigma = (P)$, be trivial. Then the Σ-singularities spectral sequence possesses a multiplicative structure. The differential d_r satisfies the following formula for every $r > 1$:*

$$d_r(a, b) = d_r(a) \cdot b + (-1)^{\deg a} a \cdot d_r(b). \ \square$$

Note 2.6.2 1) *The differential d_1, which coincides here with the Bockstein operator*

$$\beta : MG_*^\Sigma(\cdot) \longrightarrow MG_*^\Sigma(\cdot),$$

satisfies the formula

$$d_1(a,b) = d_1(a)\cdot b + (-1)^{\deg a}\frac{1}{[P]}\cdot((d_1(a)\cdot d_1(b))\cdot[\beta Q]) + (-1)^{\deg a} a\cdot d_1(b),$$

where a division by $[P]$ is justified by the structure of the algebra $E_1^{*,*}$; see section 1.6.

2) *If the admissible product structure in the bordism theory with singularities is commutative and associative then so is the product structure in the Σ-singularities spectral sequence.*

Chapter 3

The Adams-Novikov spectral sequence

Now it is high time to dwell upon the Adams-Novikov spectral sequence.

The first thing is to define all the features of the Adams-Novikov spectral sequence. Though the *homology version* is much more popular and has a nice algebraic description (see [4], [84], [108]) we are going to deal with the *cohomology version* of the Adams-Novikov spectral sequence, as it is very convenient for our purposes. Besides, we make use of the *Novikov algebraic spectral sequence*, which converges to the algebra

$$Ext^{*,*}_{A^{BP}}(BP^*(X), BP^*).$$

The latter is the second term $E_2^{*,*}$ of the Adams-Novikov spectral sequence. In addition we touch upon the results obtained by V.Vershinin and V.Gorbunov [43], [109]–[113] concerning the structure of these spectral sequences for the symplectic cobordism case.

The second thing is to apply the Adams-Novikov spectral sequence to the cobordism theory $MSp^*_\Sigma(\cdot)$ of symplectic manifolds with singularities. The purpose here is to present some of Vershinin's results [111] (Theorems 3.3.3 and 3.3.5) pertaining to a subject of interest.

The last thing is to interpret the Adams-Novikov spectral sequence for the symplectic cobordism ring from a geometric point of view. That is, we identify the above spectral sequence with the Σ-singularities spectral sequence (Theorem 3.4.1, Corollary 3.5.4). Then we formulate algebraic results concerning the Adams-Novikov spectral sequence in the geometric terms of manifolds with singularities (Corollary 3.5.5).

And finally we come to the geometric structure of the Adams-Novikov spectral sequence for the symplectic cobordism ring. This geometric description gives us the possibility of using the computations of the spectral sequences for investigating some geometric properties of the Bockstein operators so as to compute the first Adams-Novikov differential.

3.1 Basic definitions

There are certain constructions related to the Adams-Novikov spectral sequence. We deal with the *cohomology version* of the constructions, as they are much more convenient for our purposes, while the most general case is out of the scope of our interest. Anyway all the necessary cases will be considered.

Now some notes. We mean by *spectra* the objects of *stable Boardman category* [4] or of some version of it; see [10], [108]. *Ring spectra* are also required; see [4], [108]. The classifying spectrum of a multiplicative cohomology theory serves as an example of a ring spectrum.

Let h be the ring spectrum which classifies the multiplicative cohomology theory $h^*(\cdot)$. The corresponding Steenrod algebra of operations in the cohomology theory $h^*(\cdot)$ is denoted by $\mathcal{A}^h = h^*(h)$. We suppose that the algebra $\mathcal{A}^h$ is provided with the structure of a *Hopf algebra* in the standard manner; see [4], [108].

Note 3.1.1 *In the cohomology case the natural skeletal filtration provides the algebra $\mathcal{A}^h$ and the $\mathcal{A}^h$-modules $h^*(X)$ with a topology. So the cohomology theory $h^*(\cdot)$ transforms the category of spectra to the category of topologized modules over the topological algebra $\mathcal{A}^h$. We* *are not going to go into details here; for information on the homological algebra, refer to* [4], [10], [108].

Now let us describe the spectral sequence which converges to the graded set $[X, Y]_*$ of the homotopy classes of the maps between the (-1)-connected spectra. The category of spectra is quite suitable for constructing such spectral sequences. Let us consider the following exact couple in this category:

$$\begin{array}{ccccccccccc} X = X_0 & \xleftarrow{i_1} & & X_1 & \leftarrow \cdots \leftarrow & X_{n-1} & \xleftarrow{i_n} & & X_n & \leftarrow & \cdots \\ & {\scriptstyle j_0}\searrow & & \nearrow{\scriptstyle k_1} \quad \searrow & & \nearrow \quad {\scriptstyle j_{n-1}}\searrow & & \nearrow{\scriptstyle k_n} & & \searrow & \\ & & Z_1 & & \cdots & & & Z_n & & & \cdots \end{array} \tag{3.1}$$

where the maps i_q, j_q have zero degree and the map k_q has degree (-1). Applying the functor $[Y, -]_*$ to this diagram we obtain the exact couple which induces the spectral sequence converging to $[X, Y]_*$ (under some restrictions). We need some additional assumptions on the diagram (3.1) to efficiently describe this spectral sequence in terms of the given cohomology theory $h^*(\cdot)$.

Definition 3.1.1 *The diagram* (3.1) *is called the Adams resolution of the spectrum X in the cohomology theory $h^*(\cdot)$, if*

(i) *the homomorphisms $h^*(i_q)$ are trivial for all $q = 1, 2, \ldots$,*

(ii) *the induced complex*

$$h^*(X) \xleftarrow{\epsilon} h^*(Z_1) \xleftarrow{d^1} \ldots \leftarrow h^*(Z_n) \xleftarrow{d^n} h^*(Z_{n+1}) \longleftarrow \ldots$$

is the projective resolution of the A^h-module $h^*(h)$ (in the corresponding category above) where $\epsilon = h^*(j_0)$, $d^q = h^*(k_q \circ j_q)$, $q = 1, 2, \ldots$.

Note 3.1.2 *The condition* (ii) *is always true in the case when the spectra Z_n are the wedges of the spectra $\Sigma^{s_j} h$, where the sequence $\{s_j\}$ has no finite condensation points.* □

Definition 3.1.2 *The spectral sequence which is induced by the exact couple*

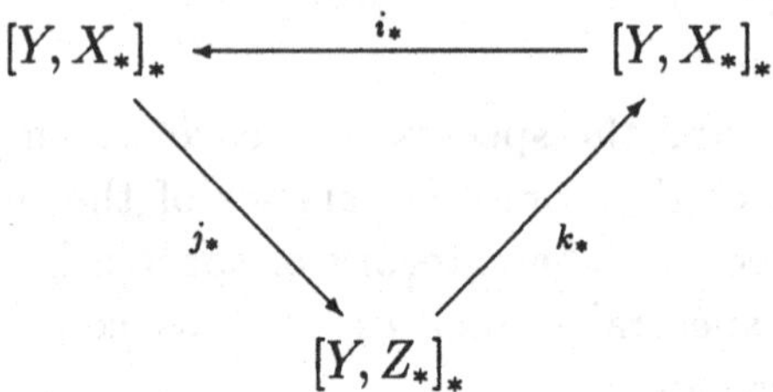

is called the Adams-Novikov spectral sequence for the spectrum X in the cohomology theory $h^*(\cdot)$ when the diagram (3.1) *is the Adams resolution.*

When it is clear what spectrum and cohomology theory are meant, the above spectral sequence is called the *Adams-Novikov spectral sequence* (*ANSS*).

According to Definition 3.1.1 we have the following isomorphism:

$$E_1^{*,*} \cong [Y, Z_*]_* \cong Hom^*_{\mathbb{A}^h}(h^*((Z_*), h^*(Y)).$$

The term $E_2^{*,*}$ coincides with the homology algebra of the complex

$$Hom^*_{\mathbb{A}^h}(h^*(Z_1), h^*(Y)) \xleftarrow{(d^1)^*} Hom^*_{\mathbb{A}^h}(h^*(Z_1), h^*(Y)) \xleftarrow{(d^2)^*} \cdots$$

So we obtain the isomorphism

$$E_2^{*,*} \cong Ext^{*,*}_{\mathbb{A}^h}(h^*(X), h^*(Y)). \tag{3.2}$$

Note 3.1.3 1) *The problems of existence of the Adams resolution and of convergence of the ANSS are not so simple; see* [4], [10], [38], [69], [77]. *When possible we are not going to touch upon them as they do have solutions in all necessary cases.*

2) *The Adams resolution may be obtained in different ways, while the corresponding spectral sequences are isomorphic beginning with the second terms $E_2^{*,*}$.*

3) *Let the spectrum X be a ring spectrum and $Y = S$ be the sphere spectrum; then the set $[S, X]_* = \pi_* X$ has the structure of a commutative ring. This ring structure is also recoverable from the Adams-Novikov spectral sequence (under some assumptions about the spectrum X) when the cohomology theory $h^*(\cdot)$ is multiplicative. Moreover the Adams-Novikov spectral sequence is a spectral sequence of commutative and associative algebras, and the product in the term $E_2^{*,*}$ coincides with the algebraic one which is determined for the functor*

$$Ext_{\mathbf{A}^h}(-, h^*);$$

see [63], [113]. *We intend to use the above multiplication structure in the cases to be considered.*

The most popular cohomology theories used for the Adams-Novikov spectral sequence are the complex cobordism theory $MU^*(\cdot)$ and its p-local version, the Brown-Peterson theory $BP^*(\cdot)$. Now let us take a fixed prime p and consider the case $h^*(\cdot) = BP^*(\cdot)$ explicitly.

We restrict to the case when $Y = S$, so that $[S, X]_* = \pi_* X$ are the homotopy groups of the spectrum X . Besides we suppose that the following condition is true:

(* *) *The spectrum X is a (-1)-connected spectrum such that for every m the group $H^m(X; \mathbf{Z}_{(p)})$ is finitely generated and torsion free.*

Theorem 3.1.3 [10], [69], [77], [84], [113] *Let the spectrum X satisfy the condition (* *); then the Adams-Novikov spectral sequence for the spectrum X in the theory $BP^*(\cdot)$ converges to the groups $\pi_* X \otimes \mathbf{Z}_{(p)}$. If the spectrum X is the ring spectrum then the Adams-Novikov spectral sequence possesses a multiplication structure.* □

We have seen already that the second term of the classic Adams-Novikov spectral sequence has the form

$$E_2^{*,*} \cong Ext^{*,*}_{\mathbf{A}^{BP}}(BP^*(X), BP^*). \tag{3.3}$$

The computation of this algebra may be rather difficult for many spectra X, because in many cases it is not possible to efficiently describe the structure of the $\mathbf{A}^{BP}$-module $BP^*(X)$.

Anyway there exist several constructions which allow us to compute the algebra (3.2). One of them is the *Novikov algebraic spectral sequence* and its generalization proposed by V.Vershinin [111].

Suppose v_k, $k = 1, 2, \ldots$, are standard Hazevinkel generators of the ring BP^*:

$$BP^* = \mathbf{Z}_{(p)}[v_1, \ldots, v_n, \ldots].$$

Let $\mathfrak{m} = (p, v_1, \ldots, v_n, \ldots)$ be the maximal prime ideal of the ring BP^*; we consider the following multiplicative filtration of the ring BP^*:

$$BP^* = \mathbf{F}_0 \supset \mathbf{F}_1 \supset \ldots \supset \mathbf{F}_n \supset \ldots, \tag{3.4}$$

where $\mathbf{F}_1 = \mathfrak{m}$, $p \in \mathbf{F}_k$ (under given $k \geq 1$) and $v_i \notin \mathbf{F}_2$ for every $i = 1, 2, \ldots$. We denote the bigraded ring associated to the ring BP^* with respect to the filtration (3.4) by $\overline{BP}^*$.

Let $\mathfrak{R}$ be a projective resolution of the A^{BP}-module $BP^*(X)$:

$$BP^*(X) \longleftarrow R_1 \longleftarrow \cdots \longleftarrow R_n \longleftarrow R_{n+1} \longleftarrow \cdots$$

The filtration (3.4) induces the filtration of the complex

$$\mathfrak{C} = Hom^*_{\mathsf{A}^{BP}}(\mathfrak{R}, BP^*)$$

as follows:

$$\mathsf{F}_n = Hom^*_{\mathsf{A}^{BP}}(\mathfrak{R}, \mathbf{F}_n). \tag{3.5}$$

As usual the filtration (3.5) determines some spectral sequence which converges to the algebra (3.3) (under some assumptions).

Definition 3.1.4 *The spectral sequence induced by the filtration* (3.5) *is called the Novikov algebraic spectral sequence (NaSS) for the spectrum* X.

It isn't difficult to determine the first term of this spectral sequence. Note that $BP^*/\mathbf{F}_1 \cong \mathbf{Z}/p$. Let $\mathcal{A}_p$ denote the ordinary mod p Steenrod algebra. Let $\mathcal{A}'_p = \mathcal{A}_p/(Q_0)$ be its quotient algebra by the two-sided ideal generated by the Bockstein operator $Q_0 \in \mathcal{A}_p$. The next isomorphism is well known (see [23]):

$$\mathsf{A}^{BP}/\left(\mathfrak{m}\cdot\mathsf{A}^{BP}\right)=\mathcal{A}'_p.$$

So the algebra $\mathcal{A}'_p$ acts naturally on the ring $\overline{BP}^*$. It may be verified directly that the graded differential module

$$\overline{\mathfrak{C}}=\sum_{n\geq 0}(\mathsf{F}_n/\mathsf{F}_{n-1})$$

is isomorphic to the module

$$Hom_{\mathsf{A}^{BP}/(\mathfrak{m}\cdot\mathsf{A}^{BP})}(\mathfrak{R}/(\mathfrak{m}\cdot\mathfrak{R}),\overline{BP}^*).$$

Note the groups $H^*(X;\mathbf{Z}_{(p)})$ are torsion free under condition (* *). In particular this means that the cohomology $H^*(X;\mathbf{Z}_{(p)})$ possesses $\mathcal{A}'_p$-module structure. The factorization by the ideal $\mathfrak{m}$ turns the projective resolution $\mathfrak{R}$ of the A^{BP}-module $BP^*(X)$ into a projective resolution of the $\mathcal{A}'$-module $\mathfrak{R}/(\mathfrak{m}\cdot\mathfrak{R})$:

$$H^*(X;\mathbf{Z}_{(p)})\longleftarrow R_1/(\mathfrak{m}\cdot R_1)\longleftarrow\cdots\longleftarrow R_n/(\mathfrak{m}\cdot R_n)\longleftarrow\cdots$$

According to the above definitions the homology of this complex coincides with the first term of the Novikov algebraic spectral sequence. We obtain the isomorphism of trigraded algebras:

$$E_1^{*,*,*}\cong Ext^{*,*}_{\mathcal{A}'_p}(H^*(X;\mathbf{Z}/(p)),\overline{BP}^*). \tag{3.6}$$

One of the statements of the next theorem is thus proved.

Theorem 3.1.5 [109], [111]–[113] *The Novikov algebraic spectral sequence is natural on the category of spectra. If the spectrum X satisfies the condition* (**), *then the spectral sequence converges to the algebra*

$$E_2^{*,*}\cong Ext^{*,*}_{\mathsf{A}^{BP}}(BP^*(X),BP^*),$$

and its initial term is isomorphic to the algebra (3.6). *If the spectrum X is the ring spectrum then the Novikov algebraic spectral sequence is the multiplicative one.* □

Note 3.1.4 *The classic Novikov algebraic spectral sequence* [77] *can be obtained by putting $k = 1$ and $\mathbf{F}_n = \mathfrak{m}^n$. We emphasize the case when $k = 2$ with the corresponding spectral sequence being called the modified Novikov algebraic spectral sequence (after V. Vershinin* [111]*). Then the abbreviations are maSS for the spectral sequence and maSS-filtration for the generating filtration.* □

The modified algebraic spectral sequence is a point of great interest. In this case the bigraded ring $\overline{BP}^*$ is isomorphic to the following polynomial ring:

$$\mathbf{Z}/p\,[h_0, h_1, \ldots, h_n, \ldots]$$

where $\deg h_0 = (2, 0)$, $\deg h_n = (-1, -2(p^n - 1))$. The algebra $\mathcal{A}'_p$ acts on the ring $\overline{BP}^*$ as follows:

$$\mathcal{P}^{p^j} h_n = \left\{ \begin{array}{ll} h_{n-1} & \textit{if } \; j = n - 1 > 0, \\ 0 & \textit{otherwise}, \end{array} \right\} \tag{3.7}$$

where $\mathcal{P}^{p^j}$ are the Steenrod powers which generate the algebra $\mathcal{A}'_p$. The formula (3.7) follows from the well known action of the Quillen algebra A^{BP}; see [109].

Note that the *maSS*-filtration of the ring $BP^* \otimes \mathbf{Z}/p$ is induced, the filtration being generated by the powers of the ideal $\tilde{\mathfrak{m}} = (v_1, \ldots, v_n, \ldots)$. The corresponding adjoined ring

$$(\mathbf{Z}/p)\,[h_1, \ldots, h_n, \ldots],$$

is denoted by $\overline{BP}^*/p$. It is also the $\mathcal{A}'_p$-module. The isomorphism

$$Ext^{*;*}_{\mathcal{A}'_p}(H^*(X; \mathbf{Z}/p), \overline{BP}^*)$$

$$\cong Ext^{*;*}_{\mathcal{A}'_p}(H^*(X; \mathbf{Z}/p), \overline{BP}^*/p) \otimes (\mathbf{Z}/p)[h_0] \tag{3.8}$$

is based on the formula (3.7). Now we are going to apply the above constructions for the symplectic cobordism case.

3.2 The modified algebraic spectral sequence

The symplectic cobordism ring MSp_* is known to have no p-primary torsion for the odd primes. We will consider the modified algebraic spectral sequence for $p = 2$. According to Theorem 3.1.5 its initial term is isomorphic to the algebra

$$Ext^{*,*}_{\mathcal{A}'_2}(H^*(MSp; \mathbf{Z}/2), \overline{BP^*}/2) \otimes (\mathbf{Z}/2)[h_0]. \tag{3.9}$$

It isn't difficult to compute the algebra (3.9); to do this we have to recall some facts on the structure of the module $H^*(MSp; \mathbf{Z}/2)$ over the Steenrod algebra $\mathcal{A}_2$.

We consider the images of the elements $Sq^{2\Delta_i} \in \mathcal{A}_2$ in the quotient algebra $\mathcal{A}'_2 = \mathcal{A}_2/(Sq^1)$ which will be denoted by $Sq^{2\Delta_i}$, where

$$\Delta_i = (\underbrace{0, \ldots, 0, 1}_{i}, 0, \ldots),$$

and the elements Sq^{Δ_i} are from the Milnor basis of the Steenrod algebra $\mathcal{A}_2$. Let $\mathcal{B}$ be the subalgebra of the algebra $\mathcal{A}'_2$ which is generated by the elements $Sq^{2\Delta_i}, i = 1, 2, \ldots$. Note the following properties:

(i) the algebra $\mathcal{B}$ is a *normal subalgebra* of the algebra $\mathcal{A}'_2$;

(ii) the quotient algebra $\mathcal{A}_2//\mathcal{B}$ is isomorphic to the quotient algebra $\mathcal{A}'_2/(Sq^2)$ of the algebra $\mathcal{A}'_2$ by a two-sided ideal generated by the operator Sq^2; see [49], [59], [76].

Let $\mathcal{A}''$ be the algebra $\mathcal{A}'//\mathcal{B} \cong \mathcal{A}'/(Sq^2)$. We use the following isomorphism (see Novikov's paper [76] for instance):

$$H^*(MSp; \mathbf{Z}/2) \cong \bigoplus_{\omega} \mathcal{A}''\sigma_\omega,$$

where σ_ω are free generators and the summing is given over all collections $\omega = (i_1, \ldots, i_q)$, where $i_t \neq 2^m - 1$, $\deg \sigma_\omega = 4(i_1 + \ldots + i_q)$.

Let us consider now the bigraded polynomial algebra

$$\mathcal{C}_* = (\mathbf{Z}/2)[c_2, \ldots, c_n, \ldots],$$

where $n = 2, 4, 5, \ldots, n \neq 2^m - 1$, $\deg c_n = (0, 0, 4n)$.

The following isomorphism is an easy consequence of the formula (3.7):

$$Ext^{*,*,*}_{\mathcal{A}'_2}(H^*(MSp; \mathbf{Z}/2), \overline{BP^*}/2) \otimes (\mathbf{Z}/2)[h_0]$$
$$\cong Ext^{*,*}_{\mathcal{A}'_2}(\mathcal{A}'', \overline{BP^*}/2) \otimes (\mathbf{Z}/2)[h_0] \otimes \mathcal{C}_*. \tag{3.10}$$

The formula (3.7) implies that $Sq^2 h = 0$, so the algebra $\mathcal{A}'_2$ acts on the ring $\overline{BP^*}$ trivially. Let the ring

$$\overline{BP}_* = (\mathbf{Z}/2)[h_0, h_1, \ldots, h_n, \ldots]$$

be dual to $\overline{BP^*}$, where $\deg h_0 = (0, 0, 2)$, $\deg h_n = (1, 0, 2(2^n - 1))$, $n = 1, 2, \ldots$.

We obtain the isomorphism

$$Ext^{*,*}_{\mathcal{A}'_2}(\mathcal{A}'_2//\mathfrak{B}, \overline{BP^*}/2) \otimes (\mathbf{Z}/2)[h_0] \otimes \mathcal{C}_*$$

$$\cong Ext^{*,*}_{\mathcal{A}'_2}(\mathcal{A}''_2, \mathbf{Z}/2) \otimes \overline{BP^*} \otimes \mathcal{C}_*.$$

Finally we need the isomorphism

$$Ext^{*,*}_{\mathcal{A}'_2}(\mathcal{A}''_2, \mathbf{Z}/2) \cong (\mathbf{Z}/2)[u_1, \ldots, u_j, \ldots], \tag{3.11}$$

where $\deg u_j = (0, 1, 2(2^j - 1))$, $j = 1, 2, \ldots$; see [109], [111], [112]. The first degree comes from the filtration's degree and is equal to zero according to the isomorphism (3.10). The isomorphisms (3.9)–(3.11) imply the proof of the following lemma.

Lemma 3.2.1 *There is a trigraded algebra isomorphism*

$$\mathsf{E}_1^{*,*,*} = Ext^{*,*}_{\mathcal{A}'_2}(H^*(MSp; \mathbf{Z}/2), \overline{BP}^*)$$

$$\cong (\mathbf{Z}/2)[c_2, \ldots, c_n, \ldots, u_1, \ldots, u_j, \ldots, h_0, h_1, \ldots, h_i, \ldots], \tag{3.12}$$

where $\deg c_n = (0, 0, 4n)$, $n = 2, 4, 5, \ldots$, $n \neq 2^m - 1$, $\deg u_j = (0, 1, 2(2^j - 1))$, $\deg h_0 = (2, 0, 0)$, $\deg h_i = (1, 0, 2(2^i - 1))$, $i, j = 1, 2, \ldots$. □

Note 3.2.1 *The proof of Lemma 3.2.1 is based on the properties of the maSS-filtration. The latter differs a little from the filtration generated by powers of the maximal proper ideal* $\mathfrak{m} = (v_0, \ldots, v_p, \ldots) \subset BP^*$. *The classic Novikov algebraic spectral sequence is determined by that filtration. Its initial term is much more complicated in comparison with the algebra* (3.12); *it was computed by V. Vershinin* [109]. □

Now we summarize the results of V.Vershinin and V.Gorbunov [21], [112] concerning the first differential of the modified algebraic spectral sequence for the symplectic cobordism ring.

But first some preliminary considerations. The Landweber-Novikov algebra $\mathcal{A}^{MSp}$ of operations in the symplectic cobordism theory has the form

$$\mathcal{A}^{MSp} \cong MSp^* \hat{\otimes} S,$$

where a free basis of the algebra S consists of the operations s_ω; see [49], [53], [77]. The Landweber-Novikov algebra $\mathcal{A}^{MSp}$ acts on all the objects which naturally depend upon the spectrum MSp and its automorphisms. In particular, it acts on the homology and cohomology groups of the spectrum MSp and on the terms E_r of every algebraic spectral sequence. The action is very important for computations of spectral sequences (see [109]–[112]), but we do not intend to describe the action of the algebra $\mathcal{A}^{MSp}$ in detail. It is sufficient to mention only some relevant facts which allow us to carefully formulate the results.

Nigel Ray determined the elements $\theta_i \in MSp_{4i-3}$ which are indecomposable for $i = 1, 2, 4, 6, \ldots$; they are of order two and are closed with respect to the action of the algebra $\mathcal{A}^{MSp}$; see [86]. The Ray elements are very important for a comprehension of the structure of the Adams-Novikov spectral sequence. Let us recall some properties of the Ray elements and consider their projection into the term $E_2^{*,*}$ of the Adams-Novikov spectral sequence and into the initial term of the modified algebraic spectral sequence.

Let φ_i be the elements θ_{2i} for $i = 1, 2, \ldots$, and the element θ_1 be a standard generator of the group $MSp_1 = \mathbf{Z}/2$. We project the element θ_1 into the term $E_2^{*,*}$ of the Adams-Novikov spectral sequence, then into the term $\mathsf{E}_1^{*,*,*}$ of the modified spectral sequence. As a result we have that the image of these projections coincides with the element u_1 for dimensional reasons. So the element u_1 is a cycle of all differentials in the *maSS* and uniquely determines (also for dimensional reasons) the element

$$u_1 \in Ext^{*,*}_{A^{BP}}(BP^*(MSp), BP^*)$$

which is also a cycle of all differentials in the *ANSS*. We consider now some Adams resolution of the spectrum MSp in the theory $BP^*(\cdot)$:

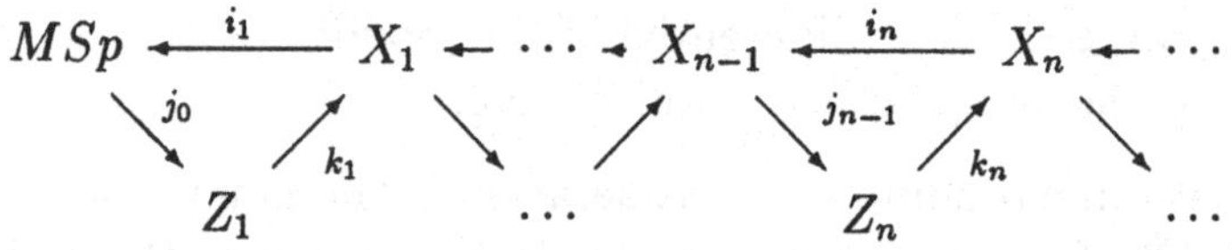

$$\begin{array}{ccccccccc} MSp & \xleftarrow{i_1} & X_1 & \leftarrow \cdots \leftarrow & X_{n-1} & \xleftarrow{i_n} & X_n & \leftarrow & \cdots \\ & \searrow^{j_0} & & \nearrow_{k_1} \quad \searrow \quad \nearrow & & \searrow^{j_{n-1}} & & \nearrow_{k_n} \quad \searrow & \\ & & Z_1 & \cdots & & & Z_n & & \cdots \end{array}$$

The corresponding Adams filtration of the ring MSp_*

$$MSp_* \longleftarrow \mathcal{F}_1 \longleftarrow \cdots \longleftarrow \mathcal{F}_{n-1} \longleftarrow \mathcal{F}_n \longleftarrow \cdots$$

where $\mathcal{F}_n = \pi_* X_n$, $n = 1, 2, \ldots$, is also invariant with respect to the action of the algebra A^{MSp}. We have seen earlier that $\theta_1 \in \mathcal{F}_1$, but $\theta_1 \notin \mathcal{F}_2$. The element φ_i lies in the group $\mathcal{F}_1$ and $\varphi_i \notin \mathcal{F}_2$ because $s_{(2i-1)}\varphi_i = \theta_1$; then $\varphi_i \notin \mathcal{F}_o \backslash \mathcal{F}_1$ since the element φ_i is of order two. We note that the line $E_\infty^{1,*}$ of the Adams-Novikov spectral sequence is monomorphically imbedded into the line $E_2^{1,*}$. So we obtain that the projection of the element φ_i into the term $E_2^{*,*}$ lies in the first line

$$Ext^{1,*}_{A^{BP}}(BP^*(MSp), BP^*).$$

The projection of the elements φ_i into the term $\mathsf{E}_1^{*,*,*}$ of the modified algebraic spectral sequence is also denoted by φ_i. For dimensional reasons it follows that the differentials cannot act on the groups $\mathsf{E}_1^{0,1,*}$. So the line $\mathsf{E}_1^{0,1,*}$ is monomorphically imbedded into the line $\mathsf{E}_\infty^{0,1,*}$, and so it is a free C_*-module with the generators $u_1, \ldots, u_j, \ldots$ according to Lemma 3.2.1, where

$$C_* = (\mathbf{Z}/2)\left[c_1, \ldots, c_n, \ldots\right],$$

where $n = 2, 4, 5, \ldots$, $n \neq 2^q - 1$. The projection of the element φ_i into the line $\mathsf{E}_1^{0,1,*}$ is also denoted by φ_i.

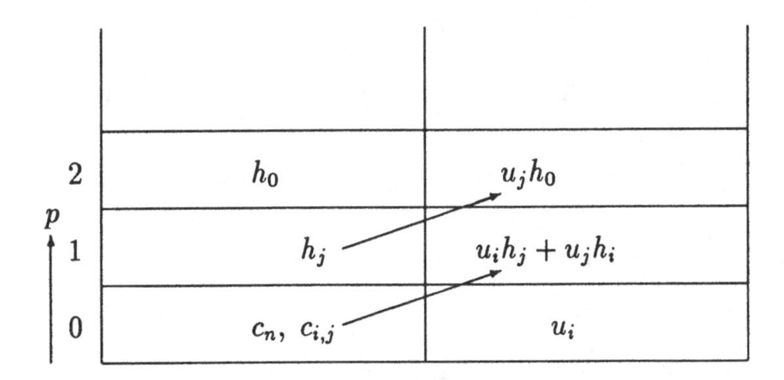

Figure 3.1: The algebra $\mathsf{E}_1^{*,*,*}$.

Note 3.2.2 *The modified algebraic spectral sequence for the symplectic cobordism ring has surprising properties. That is, the first differential is a unique nontrivial one and its action on the generators of the algebra* $\mathsf{E}_1^{*,*,*}$ *depends only on the binary decomposition of their indexes.* □

Let us introduce the following notation.

Let $c_{i,j}$ denote c_n if $n = 2^{i-1} + 2^{j-1} - 1$, $1 \leq i < j$. In the case $n = 2m - 1$ where $m = 2^{i_1 - 2} + \ldots + 2^{i_q - 2}$ is a binary decomposition of the number m (here $2 \leq i_1 < \ldots < i_q$, $q \geq 3$), the element c_n is denoted by $c_{i_1,\ldots,i_q}$. The corresponding Ray elements are denoted by $\varphi_{i_1,\ldots,i_q}$, and their projection to the term $\mathsf{E}_1^{*,*,*}$ by $\Phi_{i_1,\ldots,i_q}$.

The following theorem describes the action of the first differential d_1^{Al} of the modified algebraic spectral sequence for MSp_*.

Theorem 3.2.2 *There exist polynomial generators of the algebra*

$$\mathsf{E}_1^{*,*,*} \cong \mathbf{Z}/2\,[c_2, \ldots, c_n, \ldots, u_1, \ldots, u_j, \ldots, h_0, h_1, \ldots, h_i, \ldots],$$

such that the following statements are true.

1° *The differential* d_1^{Al} *acts on the generators* c_n *as follows:*

(a) *if n is even and is not a power of 2, then the element c_n is a cycle of all differentials of the modified algebraic spectral sequence;*

(b) *if* $n = 2^{i-1} + 2^{j-1} - 1$, $1 \le i < j$, *then*

$$d_1^{Al}(c_{i,j}) = u_i h_j + u_j h_i; \tag{3.13}$$

(c) *if* $n = 2^{i_1-1} + \ldots + 2^{i_q-1} - 1$, $2 \le i_1 < \ldots < i_q$, $q \ge 3$, *then*

$$d_1^{Al}(c_{i_1,\ldots,i_q}) = \sum_{1\le s<t\le q} (u_{i_t} h_{i_s} + u_{i_s} h_{i_t}) c_{1,i_1} \ldots \widehat{c}_{1,i_s} \ldots \widehat{c}_{1,i_t} \ldots c_{1,i_q}; \tag{3.14}$$

(d) *there is an isomorphism* $\mathsf{E}_2^{*,*,0} \cong \mathsf{E}_\infty^{*,*,0}$.

2° *The generators* u_j *are the cycles of all differentials, and the following equalities are true in the algebra* $\mathsf{E}_1^{*,*,*}$:

$$u_1 = \theta_1,\ u_i = \Phi_{2^i-2},\quad i = 2, 3, \ldots,$$
$$\Phi_{i,j} = u_i c_{1,j} + u_j c_{1,i} + \sum_J c_J \varphi_J,\ 1 < i < j,$$

where $\Phi_{i,j}$ *is a projection of* $\varphi_{2^{i-1}+2^{j-1}}$, *and if* $q \ge 3$ *the Ray element* $\varphi_{i_1,\ldots,i_q}$ *projects to*

$$\Phi_{i_1,\ldots,i_q} = u_1 c_{i_1,\ldots,i_q} + \sum_{i=1}^{q} u_{i_t} c_{1,i_1} \ldots \widehat{c}_{1,i_t} \ldots c_{1,i_q} + \sum_J \Phi_J c_J, \tag{3.15}$$

where the elements $c_J \in \mathsf{E}_1^{0,0,*}$ *above are the cycles of all the differentials.*

3° *The formula*

$$d_1^{Al} h_j = h_0 u_j. \tag{3.16}$$

holds for every $j = 1, 2, \ldots$ □

Note 3.2.3 *The statements* 1°(a), 1°(d), 3° *were proved by V. Vershinin* [109], [110], *the other by V. Vershinin & V. Gorbunov* [21], [112]. *These proofs are essentially based on the known module structure over the Landweber-Novikov algebra as well as on Buchstaber's results concerning Two-valued Formal Group Theory (the statements* 1°(a), 1°(d)*).* □

The algebra $\mathrm{E}_1^{*,*,*}$ and the action of the first differential d_1^{Al} are shown in Figure 3.1.

Here the statements of Theorem 3.2.2 are cited in their original form. For our purposes it is convenient to take some other generators $c_{i_1,\dots,i_q}$ for $q \geq 3$.

Lemma 3.2.3 *There exist generators $c_{i_1,\dots,i_q}$, where $q \geq 3$, such that the following formulas are true:*

$$d_1^{Al}(c_{i_1,\dots,i_q}) = u_1 \sum_{t=1}^{q} h_{i_t} c_{i_1,\dots,\widehat{i_1},\dots,i_q} + h_1 \sum_{t=1}^{q} u_{i_t} c_{i_1,\dots,\widehat{i_1},\dots,i_q} \tag{3.17}$$

where the elements $c_{i,j}$ coincide with the generators from point 1^0(b) *of Theorem 3.2.2.*

Proof. Let us apply induction with respect to $q \geq 3$. We put

$$c'_{i,j,k} = c_{i,j}c_{k,1} + c_{i,k}c_{j,1} + c_{j,k}c_{i,1} + c_{i,j,k},$$

for the element $c_{i,j,k}$ taken from the formula (3.17), where $2 \leq i < j < k$. Hence we have

$$d_1^{Al} c'_{i,j,k} = h_1(u_k c_{i,j} + u_j c_{i,k} + u_i c_{j,k}) + u_1(h_k c_{i,j} + h_j c_{i,k} + h_i c_{j,k}).$$

Let us make the induction step. We write

$$Z^{(1)}_{i_1,\dots,i_n} = \sum_{1 \leq s < t \leq n} c_{i_s,i_t} c_{1,i_1} \dots \widehat{c}_{1,i_s} \dots \widehat{c}_{1,i_t} \dots c_{1,i_n}$$

for all n and define the following elements:

$$Z^{(k)}_{i_1,\dots,i_n} \sum_{1 \leq t_1 < \dots < t_k \leq n} c_{i_{t_1},\dots,i_{t_k}} c_{1,i_1} \dots \widehat{c}_{1,i_{t_1}} \dots \widehat{c}_{1,i_{t_k}} \dots c_{1,i_n}$$

for all $n \leq q$, $k = 1, 2, \dots$. It is clear that

$$d_1^{Al}\left(c_{i_1,\dots,i_{q+1}} + \sum_{k=1}^{q-1} Z^{(k)}_{i_1,\dots,i_q} \right)$$

$$= u_1 \sum_{t=1}^{q} h_{i_t} c'_{i_1,\dots,\widehat{i_1},\dots,i_q} + h_1 \sum_{t=1}^{q} u_{i_t} c'_{i_1,\dots,\widehat{i_1},\dots,i_q}. \quad \square$$

Later we'll use the chosen generators $c_{i_1,\dots,i_q}$ from Lemma 3.2.3.

3.3 Symplectic cobordism with singularities

Here we will discuss some results obtained by V.Vershinin [111]. For clarity we need to present some of them together with their proofs.

Let us consider the following sequence of closed symplectic manifolds:

$$\Sigma = (P_1, \ldots, P_k, \ldots),$$

where $[P_1] = \theta_1$, $[P_k] = \varphi_{2^k-2}$ for $k \geq 2$, φ_j being the Ray elements. As before we make use of the notation $\Sigma_k = (P_1, \ldots, P_k)$ for $k = 1, 2, \ldots$. Let $MSp_*^{\Sigma}(\cdot)$, $MSp_*^{\Sigma_k}(\cdot)$ be the corresponding bordism theories with singularities and MSp^{Σ}, MSp^{Σ_k} their classifying spectra respectively. Sometimes the spectrum MSp can be alternatively denoted by MSp^{Σ_0}. Note that the integer cohomology groups of the spectrum MSp are nonzero only in dimensions $\equiv 0 \bmod 4$, and are torsion free.

Lemma 3.3.1 *The cohomology groups*

$$H^*(MSp^{\Sigma_k}; \mathbf{Z})$$

are nontrivial only in even dimensions, are finitely generated, and are torsion free for every $k = 1, 2, \ldots$.

Proof. The Bockstein-Sullivan triangle (1.2) induces the cofibration:

$$\Sigma^{p_k} MSp^{\Sigma_{k-1}} \xrightarrow{\times [P_k]} MSp^{\Sigma_{k-1}} \xrightarrow{\pi_{k-1}^k} MSp^{\Sigma_k} \tag{3.18}$$

where $p_k = \dim P_k = 2^{k+1} - 3$. Induction with respect to k allows us to conclude that corresponding cohomology exact sequences (with integer and mod p coefficients) split into the following short ones:

$$0 \leftarrow H^m(MSp^{\Sigma_{k-1}}) \xleftarrow{(\pi_k^{k-1})^*} H^m(MSp^{\Sigma_k}) \xleftarrow{\delta_k^*} H^{m-p_k-1}(MSp^{\Sigma_{k-1}}) \leftarrow 0$$

These short exact sequences immediately imply the desired statement. □

The spectrum MSp^{Σ_k} satisfies the condition (* *) from section 3.1 for every k, hence there exists a corresponding Adams-Novikov spectral sequence as well as a modified algebraic spectral sequence. Now let us consider the case when $p = 2$ and the corresponding Brown-Peterson cohomology theory $BP^*(\cdot)$. The odd prime case is trivial due to the following lemma.

Lemma 3.3.2 *The groups $MSp_*^{\Sigma_k}$ are p-torsion free for all $k = 1, 2, \dots$ and every odd prime p.*

Proof. Induction with respect to k. We begin by recalling the fact that the Bockstein-Sullivan exact sequence

$$\dots \to MSp_*^{\Sigma_{k-1}} \xrightarrow{\times [P_k]} MSp_*^{\Sigma_{k-1}} \xrightarrow{\pi_{k-1}^k} MSp_*^{\Sigma_k} \xrightarrow{\delta_k} MSp_*^{\Sigma_{k-1}} \to \dots$$

tensor multiplied on $\mathbf{Z}_{(p)}$ (where $p > 2$) splits into

$$0 \to MSp_*^{\Sigma_{k-1}} \otimes \mathbf{Z}_{(p)} \xrightarrow{\times [P_k]} MSp_*^{\Sigma_{k-1}} \otimes \mathbf{Z}_{(p)} \xrightarrow{\pi_{k-1}^k} MSp_*^{\Sigma_k} \otimes \mathbf{Z}_{(p)} \to 0$$

The induction assumption implies that the groups $MSp_*^{\Sigma_k}$ are p-torsion free for all $p > 2$. □

The morphism of the initial terms of the modified algebraic spectral sequences for $p = 2$,

$$\pi_*^k : \mathsf{E}_1^{*,*,*}(MSp) \longrightarrow \mathsf{E}_1^{*,*,*}(MSp^{\Sigma_k}),$$

is induced by the spectrum map

$$\pi^k : MSp \longrightarrow MSp^{\Sigma_k}.$$

The algebra $\mathsf{E}_1^{*,*,*}(MSp)$ is a polynomial algebra over $\mathbf{Z}/2$ with the generators c_n, u_i, h_j. The images of these generators in the algebras $\mathsf{E}_1^{*,*,*}(MSp^{\Sigma_k})$ are also denoted by c_n, u_i, h_j.

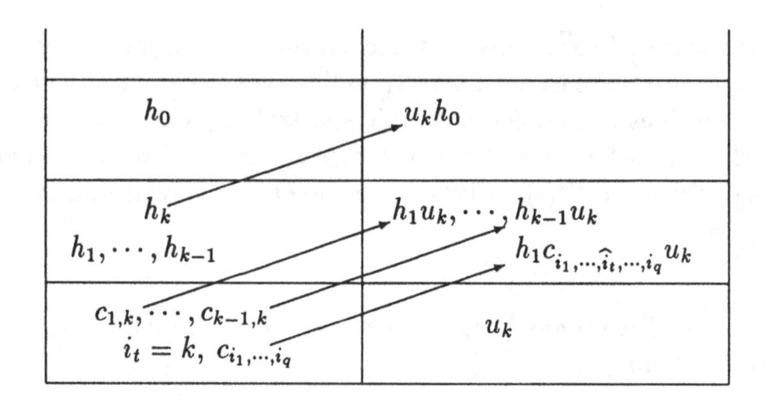

Figure 3.2: The algebra $\mathsf{E}_1^{*,*,*}(MSp^{\Sigma_{k-1}})$

Theorem 3.3.3 *There exists a commutative and associative admissible product structure* μ_k *in the symplectic bordism theory with the singularities* $MSp_*^{\Sigma_k}(\cdot)$ *such that the coefficient ring* $MSp_*^{\Sigma_k}$ *is isomorphic to the following polynomial ring up to dimension* $2^{k+2}-3$*:*

$$\mathbf{Z}\,[w_1, \dots, w_k, \dots, x_2, \dots, x_n, \dots\,],$$

where the elements w_k *can be presented by the symplectic manifolds* W_k *such that* $\partial W_k = 2P_k$, $\deg w_i = 2(2^i - 1)$, $i = 1, 2, \dots$, $n = 2, 4, 5, \dots$, $n \neq 2^m - 1$. *The initial term of the modified algebraic spectral sequence for the spectrum* MSp^{Σ_k} *for* $k \geq 1$ *is isomorphic to the algebra*

$$(\mathbf{Z}/2)\,[c_2, \dots, c_n, \dots, u_{k+1}, \dots, u_j, \dots, h_0, h_1, \dots, h_i, \dots]. \qquad (3.19)$$

In addition, $\pi_*^k(u_q) = 0$ *for* $q = 1, \dots, k$, $u_q \in \mathsf{E}_1^{*,*,*}(MSp)$.

Proof. Induction with respect to k We assume that all the statements are proved for the bordism theory $MSp_*^{\Sigma_{k-1}}(\cdot)$. Now let us consider the *maSS* for $MSp^{\Sigma_{k-1}}$ more explicitly. The algebra $\mathsf{E}_1^{*,*,*}(MSp^{\Sigma_{k-1}})$ is shown in Figure 3.2 in homotopy dimensions up to $2^{k+2}-3$.

Figure 3.2 shows the generators and the action of the first differential. This action is determined by the formulas for the differential d_1^{Al} (Theorem 3.2.2). The induction assumption implies that the spectrum $MSp^{\Sigma_{k-1}}$ is a ring spectrum, so the spectral sequence possesses multiplicative properties.

So the cycles (of given dimensions) of all the differentials lying in the algebra

$$\mathsf{E}_1^{s,*,*}(MSp^{\Sigma_{k-1}}), \quad s \geq 1,$$

have the form cu_k, where $c \in \mathsf{E}_1^{0,*,*}(MSp^{\Sigma_{k-1}})$ is a cycle as well.

Hence elements (of the same homotopy dimension) of the first line $E_2^{1,*}(MSp^{\Sigma_{k-1}})$ of the Adams-Novikov spectral sequence have the form $y \cdot \varphi_{2^k-2}$, where $y \in E_2^{0,*}$. The *ANSS* possesses a product structure, so we have

Lemma 3.3.4 *The torsion elements of Tors $MSp_*^{\Sigma_k}$ of dimensions up to $2^{k+2}-3$ have the form $z \cdot \varphi_{2^k-2}$, where $z \in MSp_*^{\Sigma_k}$.* □

Note 3.3.1 *The above arguments are true when $k \geq 2$; in the case $k = 1$, the statement of the lemma is obvious.* □

Now we are going to prove the existence of an admissible product structure. Here the results of Chapter 2 would come in handy. The obstruction to existence of the admissible product structure μ_k is of dimension $2^{k+2}-5$. The Ray element φ_{2^k-2} is of order two, so the manifold-obstruction P_k' is Σ_k-bordant to a manifold without singularities. In particular, we have

$$[P_k']_{\Sigma_k} \in \operatorname{Im}\left(\pi_k^{k-1} : MSp_*^{\Sigma_k} \longrightarrow MSp_*^{\Sigma_{k-1}}\right).$$

According to Lemma 3.3.4 every preimage of the element $[P_k']_{\Sigma_k}$ in the group $MSp_*^{\Sigma_{k-1}}$ has the form $z \cdot \varphi_{2^k-2}$. Hence $[P_k']_{\Sigma_k} = 0$ due to the Bockstein-Sullivan exact sequence.

Now we prove commutativity of the product structure μ_k, which is determined by the product structure μ_{k-1} in the theory $MSp_*^{\Sigma_{k-1}}(\cdot)$ and by some Σ_k-manifold Q_k which bounds the obstruction P_k'. An obstruction to commutativity is the Σ_k-manifold

$$B_k = Q_k \cup_\tau -Q_k;$$

see section 2.3. Suppose $\mathbf{b}_k = [B_k]_{\Sigma_k}$. According to Lemma 2.3.2 the element $\mathbf{b}_k$ is of order two in the group $MSp_{2^{k+2}-4}^{\Sigma_k}$. Let us consider the

Bockstein-Sullivan exact sequence

$$\dots \xrightarrow{\cdot\varphi_{2k-2}} MSp^{\Sigma_{k-1}}_{2^{k+2}-4} \xrightarrow{\pi^k_{k-1}} MSp^{\Sigma_k}_{2^{k+2}-4} \xrightarrow{\delta_k} MSp^{\Sigma_{k-1}}_{2^{k+1}-8} \xrightarrow{\cdot\varphi_{2k-2}} \dots$$

The group $MSp^{\Sigma_{k-1}}_{2^{k+1}-8}$ is torsion free, so $\mathbf{b}_k \in \operatorname{Im} \pi^{k-1}_k$. Let $\widehat{b}_k$ be some preimage of the element $\mathbf{b}_k$. Lemma 3.3.4 implies that the element $\widehat{b}_k$ lies in the image of the homomorphism $\cdot\varphi_{2k-2}$. Hence $\mathbf{b}_k = 0$ due to exactness, so the product structure μ_k is commutative (Theorem 2.3.1). Associativity of the product structure μ_k is a consequence of Lemma 2.4.2 since the corresponding obstruction is of order three.

Now let us prove the statement concerning the structure of the initial term of the modified algebraic spectral sequence. Lemma 3.3.1 implies the following short exact sequence with coefficients in $\mathbf{Z}/2$:

$$0 \leftarrow H^m(MSp^{\Sigma_{k-1}}) \xleftarrow{(\pi^{k-1}_k)^*} H^m(MSp^{\Sigma_k}) \xleftarrow{\delta^*_k} H^{m-p_k-1}(MSp^{\Sigma_{k-1}}) \leftarrow 0$$

The sequence can be considered as an exact sequence of $\mathcal{A}'_2$-modules. Note that it corresponds to the cofibration (3.19). We apply the functor

$$Ext_{\mathcal{A}'_2}\left(-, \overline{BP^*}\right)$$

to this exact sequence to obtain the following exact triangle:

$$\begin{array}{ccc} \mathsf{E}_1^{*,*,*}(MSp^{\Sigma_{k-1}}) & \xleftarrow{\omega_k} & \mathsf{E}_1^{*,*,*}(MSp^{\Sigma_{k-1}}) \\ & {\scriptstyle (\pi^{k-1}_k)_*} \searrow \quad \nearrow {\scriptstyle (\delta_k)_*} & \\ & \mathsf{E}_1^{*,*,*}(MSp^{\Sigma_k}) & \end{array} \tag{3.20}$$

where ω_k is a connecting homomorphism. The natural properties of *maSS* imply that the homomorphism ω_k is adjoined to the homomorphism generated by a multiplication by the element $\varphi_{2k-2} \in MSp^{\Sigma_{k-1}}_*$. Thus the homomorphism ω_k coincides with a multiplication by the element u_k being the projection of the element φ_{2k-2}. According to the induction assumptions the element u_k is a generator of the polynomial algebra. So the triangle (3.20) implies the isomorphism

$$\mathsf{E}_1^{0,*,*}(MSp^{\Sigma_k}) \cong \mathsf{E}_1^{0,*,*}(MSp^{\Sigma_{k-1}})/\mathsf{E}_1^{0,*,*}(MSp^{\Sigma_{k-1}}) \cdot u_k.$$

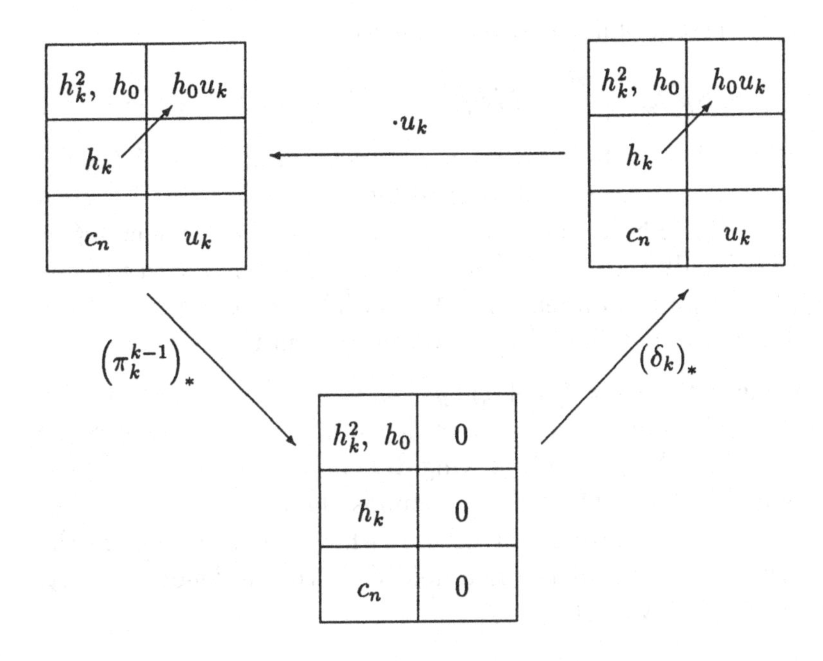

Figure 3.3: The triangle (3.20)

In particular we can see that the groups $\mathsf{E}_1^{*,s,*}(MSp^{\Sigma_k})$ are zero for $s \geq 1$ for homotopy dimensions up to $2^{k+2}-3$. It is obvious that the cells of the line $E_2^{s,*}$ of the Adams-Novikov spectral sequence are also trivial in the same dimensions for all $s \geq 1$. So the ring $MSp_*^{\Sigma_k}$ is torsion free in the same dimensions.

Finally, we have to prove the polynomiality of the ring $MSp_*^{\Sigma_k}$ in dimensions up to $2^{k+2}-3$. We note that the ring homomorphism

$$\pi_k^{k-1} : MSp_*^{\Sigma_{k-1}} \longrightarrow MSp_*^{\Sigma_k}$$

is an isomorphism in dimensions up to $2^{k+1}-3$ since $\dim P_k = 2^{k+1}-3$.

Now let us consider carefully the triangle (3.20); see Figure 3.3. We have the isomorphism

$$E_2^{*,*,*} \cong E_\infty^{*,*,*}$$

in dimension $2(2^k-1)$. At the level of the terms $\mathsf{E}_\infty^{*,*,*}$ we get

$$(\delta_k)_* h_k = h_0.$$

Due to the Bockstein-Sullivan exact sequence

$$\cdots \xrightarrow{\cdot\varphi_{2k-2}} MSp^{\Sigma_{k-1}}_{2(2^k-1)} \xrightarrow{\pi^k_{k-1}} MSp^{\Sigma_k}_{2(2^k-1)} \xrightarrow{\delta_k} MSp^{\Sigma_{k-1}}_0 \xrightarrow{\cdot\varphi_{2k-2}} \cdots$$

the element w_k adjoined to the element h_k has to be such that $\delta_k(w_k) = 2$. The element w_k may be determined by an Sp-manifold W_k, such that $\partial W_k = 2P_k$. The element h_k^2 is adjoined to the element $w_k^2 = \mu(w_k \otimes w_k)$ by multiplicativity. This proves the polynomiality of the ring $MSp^{\Sigma_k}_* \otimes \mathbf{Z}_{(2)}$ up to dimension $2^{k+2} - 3$. The last thing is that the element w_k^2 cannot be divided by an odd number in the ring $MSp^{\Sigma_k}_*$.

We assume that $w_k^2 = (4q + \epsilon) \cdot y$, where $\epsilon = \pm 1$. Theorem 2.2.4 allows us to make some correction to the product μ_k. Suppose the manifold S is some Σ_k-manifold having dimension $\dim S_k = 2^{k+2} - 4$; then the manifold $Q_k^0 = Q_k \cup S$ also bounds the obstruction P_k'. Let $[S]_{\Sigma_k} = q \cdot y$. The new product structure (which is determined by the product structure μ_{k-1} and the manifold Q_k^0) will be denoted by $\tilde{\mu}_k$. Due to formula (2.9) we get

$$\begin{aligned}\tilde{\mu}_k(w_k \otimes w_k) &= \mu_k(w_k \otimes w_k) - \mu_k(\mu_k(\beta_k w_k \otimes \beta_k w_k) \otimes q \cdot y) \\ &= (4q + \epsilon) \cdot y - 4q \cdot y = \epsilon \cdot y.\end{aligned}$$

So we have the product structure $\tilde{\mu}_k$, such that the ring $MSp^{\Sigma_k}_*$ is polynomial up to dimension $2^{k+2} - 3$. The theorem is proved. □

Note 3.3.2 *We emphasize that the product structure $\tilde{\mu}_k$ does exist for every manifold W_k and is always commutative and associative.* □

Taking a limit over k we get the following result.

Theorem 3.3.5 *There exists a commutative and associative admissible product structure μ in the symplectic bordism theory with singularities $MSp^{\Sigma}_*(\cdot)$ such that the coefficient ring MSp^{Σ}_* coincides with the polynomial ring:*

$$MSp^{\Sigma}_* \cong \mathbf{Z}[w_1, \ldots, w_k, \ldots, x_2, \ldots, x_n, \ldots], \qquad (3.21)$$

where the generators w_k may be presented by any Sp-manifolds W_k, such that $\partial W_k = 2P_k$, $\deg w_k = 2(2^k - 1)$, $k = 1, 2, \ldots$, $\deg x_n = 4n$, $n = 2, 4, 5, \ldots$, $n \neq 2^m - 1$. □

The cohomology groups of the spectra MSp^{Σ_k} and MSp^{Σ} can also be computed. As we have seen earlier the cohomology groups

$$H^*(MSp^{\Sigma_k}; \mathbf{Z}/p), \quad H^*(MSp^{\Sigma}; \mathbf{Z}/p)$$

are free modules over the algebra $\mathcal{A}'_p = \mathcal{A}_p/(Q_0)$ for $p > 2$.

We consider the case $p = 2$. Let $\mathcal{B}_k$ be the subalgebra of the Steenrod algebra $\mathcal{A}_2$, whose basis consists of the elements Sq^J, where $J = \{i_1, \ldots, i_n, \ldots\}$ are *admissible sequences* such that $i_t \le 1$, if $t \le k$, and $i_t \ge 3$, if $t > k$. It isn't difficult to prove that $\mathcal{B}_k$ is a normal subalgebra of the algebra $\mathcal{A}_2$ for each k.

The following theorem was proved by V.Vershinin; see [111].

Theorem 3.3.6 *The* $(\mathbf{Z}/2)$*-cohomology of the spectra* MSp^{Σ_k} *is a free module over the algebra* $\mathcal{A}_2//\mathcal{B}_k$ *for every* k *and that of the spectrum* MSp^{Σ} *is a free module over the algebra* $\mathcal{A}'_2$. □

Finally we consider the cohomology ring $H_*(MSp^{\Sigma}; \mathbf{Z})$.

The groups $H_*(MU; \mathbf{Z})$ are computed by means of Chern classes or Formal Group Theory. In our case it is simpler to apply the classic mod p Adams spectral sequence for the spectrum $MSp^{\Sigma} \wedge K(\mathbf{Z})$. Following the speculations of [76], we have the following result.

Corollary 3.3.7 *There is an isomorphism*

$$H_*(MSp^{\Sigma}; \mathbf{Z}) \cong \mathbf{Z}\left[m_1, \ldots, m_k, \ldots, c_2, \ldots, c_n, \ldots\right],$$

where $\deg\ m_k = 2(2^k - 1)$, $k = 1, 2, \ldots$, $\deg\ c_n = 4n$, $n = 2, 4, 5, \ldots$, $n \neq 2^m - 1$. □

Now we can come back to the Adams-Novikov spectral sequence for MSp_*.

3.4 The Σ-SSS for MSp

The above considerations allow us to conclude that the bordism theory with singularities $MSp_*^{\Sigma}(\cdot)$ is quite suitable for constructing the Adams-Novikov spectral sequence. Now we intend to present a particular *Adams resolution* for the spectrum MSp.

Let us consider a corresponding Σ-singularities spectral sequence (it is called the Σ-SSS for symplectic cobordism). According to section 1.6, we have the following exact couple:

$$\begin{array}{ccccccccccc}
MSp_*(\cdot) & \xleftarrow{\gamma(1)} & MSp_*^{\Sigma\Gamma(1)}(\cdot) & \leftarrow \cdots \leftarrow & MSp_*^{\Sigma\Gamma(n-1)}(\cdot) & \xleftarrow{\gamma(n)} & MSp_*^{\Sigma\Gamma(n)}(\cdot) & \leftarrow \cdots \\
& {\scriptstyle\pi(0)}\searrow & & \nearrow{\scriptstyle\partial_1} \quad \searrow \quad \nearrow & & {\scriptstyle\pi(n-1)}\searrow & & \nearrow{\scriptstyle\partial_n} \quad \searrow \\
& & MSp_*^{\Sigma(1)}(\cdot) & \cdots & & & MSp_*^{\Sigma(n)}(\cdot) & & \cdots
\end{array}$$

The diagram determines the following diagram of classifying spectra:

$$\begin{array}{ccccccccccc}
MSp & \xleftarrow{\gamma(1)} & MSp^{\Sigma\Gamma(1)} & \leftarrow \cdots \leftarrow & MSp^{\Sigma\Gamma(n-1)} & \xleftarrow{\gamma(n)} & MSp^{\Sigma\Gamma(n)} & \leftarrow \cdots \\
& {\scriptstyle\pi(0)}\searrow & & \nearrow{\scriptstyle\partial_1} \quad \searrow \quad \nearrow & & {\scriptstyle\pi(n-1)}\searrow & & \nearrow{\scriptstyle\partial_n} \quad \searrow \\
& & MSp^{\Sigma(1)} & \cdots & & & MSp^{\Sigma(n)} & & \cdots
\end{array} \tag{3.22}$$

Theorem 3.4.1 *The diagram* (3.22) *is an Adams resolution of the spectrum MSp in the cohomology theory $MSp_{\Sigma}^*(\cdot)$.*

Proof. The above definition dictates the necessity of proving the following statements:

(i) *the homomorphism $MSp_{\Sigma}^*(\gamma(k))$ is trivial for every k;*

(ii) *the complex*

$$MSp_{\Sigma}^*(MSp) \xleftarrow{\epsilon_0} MSp_{\Sigma}^*(MSp^{\Sigma(1)}) \xleftarrow{d^1} \cdots \leftarrow MSp_{\Sigma}^*(MSp^{\Sigma(n)}) \xleftarrow{d^n} \cdots \tag{3.23}$$

is a projective resolution of the A^{Σ}-module $MSp_{\Sigma}^*(MSp)$, where $\epsilon_j = MSp_{\Sigma}^*(\pi(0))$, $d^k = MSp_{\Sigma}^*(\pi(k)\circ\partial(k))$, $k = 1, 2, \ldots$.

Here A^{Σ} is the operation algebra in the cohomology theory $MSp^*_{\Sigma}(\cdot)$. It should be mentioned that the structure of this algebra is unknown. The spectra $MSp^{\Sigma(k)}$ split into the wedges of suspensions of the spectrum MSp^{Σ}. So the A^{Σ}-modules $MSp^*_{\Sigma}(MSp^{\Sigma(k)})$ are free for all k. The dimensions of the manifolds P_k increase: $\dim P_k = 2^{k+2} - 3$. So it is sufficient to prove exactness of the complex (3.23).

To do this we should examine the cohomology homomorphisms, which are induced by the maps $\gamma(k)$, $\pi(k)$ and $\partial(k)$.

We will use the notation $MSp^{\Sigma_k}_*(\cdot)$, where $\Sigma_k = (P_1, \ldots, P_k)$ for every $k = 1, 2, \ldots$ as above.

We remark that there are defined Bockstein operators acting in every bordism theory $MSp^{\Sigma_k}_*(\cdot)$,

$$\beta_j^{(k)} : MSp^{\Sigma_k}_*(\cdot) \longrightarrow MSp^{\Sigma_k}_*(\cdot), \quad k = 1, 2, \ldots,$$

which uniquely determine the maps of the classifying spectra,

$$\beta_j^{(k)} : MSp^{\Sigma_k} \longrightarrow MSp^{\Sigma_k}$$

up to homotopy. The corresponding homomorphisms in integral cohomology are

$$(\beta_j^{(k)})^* : H^*(MSp^{\Sigma_k}) \longrightarrow H^*(MSp^{\Sigma_k}).$$

It is clear that the homomorphisms $(\beta_j^{(k)})^*$ satisfy the following equalities (see section 1.3):

$$(\beta_j^{(k)})^* \circ (\beta_j^{(k)})^* = 0, \quad (\beta_i^{(k)})^* \circ (\beta_j^{(k)})^* = (\beta_j^{(k)})^* \circ (\beta_i^{(k)})^*. \tag{3.24}$$

In particular,

$$\text{Im } (\beta_j^{(k)})^* \subseteq \text{Ker } (\beta_j^{(k)})^*.$$

Lemma 3.4.2 *There is the inclusion*

$$\text{Im } (\beta_j^{(k)})^* \supseteq \text{Ker } (\beta_j^{(k)})^*$$

for every k, j, $1 \leq j \leq k$.

Proof. 1^0 Let $j = k$. Then the map $\beta_k^{(k)}$ is homotopic to the composition

$$MSp^{\Sigma_k} \xrightarrow{\delta_k} \Sigma^{p_k+1} MSp^{\Sigma_{k-1}} \xrightarrow{\pi_k^{k-1}} \Sigma^{p_k+1} MSp^{\Sigma_k}.$$

From the corresponding cohomology exact sequence (Lemma 3.3.1) it follows that $(\pi_k^{k-1})^*$ is an epimorphism and δ_k^* is a monomorphism. So we get the equalities

$$\mathrm{Ker}\ (\beta_k^{(k)})^* = \mathrm{Ker}\ (\delta_k)^* = \mathrm{Im}\ (\beta_k^{(k)})^*.$$

2^0 The lemma is supposed to be proved for all spectra MSp^{Σ_n} when $n \leq k-1$. The statement is proved in 1^0 for $j = k$. Let $j < k$. Now the point of departure is the diagram

$$\begin{array}{ccccc}
H^*(MSp^{\Sigma_k}) & \xrightarrow{(\beta_j^{(k)})^*} & H^*(MSp^{\Sigma_k}) & \xrightarrow{(\beta_j^{(k)})^*} & H^*(MSp^{\Sigma_k}) \\
\downarrow (\pi_j^{(k)})^* & & \downarrow (\pi_j^{(k)})^* & & \downarrow (\pi_j^{(k)})^* \\
H^*(MSp^{\Sigma_j}) & \xrightarrow{(\beta_j^{(j)})^*} & H^*(MSp^{\Sigma_j}) & \xrightarrow{(\beta_j^{(j)})^*} & H^*(MSp^{\Sigma_j}) \\
\downarrow & & \downarrow & & \downarrow \\
0 & & 0 & & 0
\end{array} \tag{3.25}$$

Let $x \in \mathrm{Ker}\ (\beta_j^{(k)})^* \subset H^*(MSp^{\Sigma_k})$. The induction implies that there exists the element $y \in H^*(MSp^{\Sigma_k})$, such that

$$-(\beta_j^{(k)})^*(y) + x \in \ \mathrm{Ker}\ (\pi_j^{(k)})^*.$$

Also if $x_1 = -(\beta_j^{(k)})^*(y) + x$, then

$$(\beta_j^{(k)})^*(x_1) = -(\beta_j^{(k)})^* \circ (\beta_j^{(k)})^*(y) + (\beta_j^{(k)})^*(x) = 0$$

since $(\beta_j^{(k)})^* \circ (\beta_j^{(k)})^* = 0$. The induction implies that there exist the elements $x_i, y_i \in H^*(MSp^{\Sigma_k})$, such that

a) $x_i = -(\beta_j^{(k)})^*(y_i) + x_i \in \operatorname{Ker}(\pi_{j+1}^{(k)})^*$,

b) $(\beta_j^{(k)})^*(y_i) = 0$,

c) $x_0 = x_1, y_0 = y_1$.

It is convenient to proceed with the diagram

$$
\begin{array}{ccccc}
0 & & 0 & & 0 \\
\downarrow & & \downarrow & & \downarrow \\
H^*(MSp^{\Sigma_{k-1}}) & \xrightarrow{(\beta_j^{(k-1)})^*} & H^*(MSp^{\Sigma_{k-1}}) & \xrightarrow{(\beta_j^{(k-1)})^*} & H^*(MSp^{\Sigma_{k-1}}) \\
\delta_k^* \downarrow & & \delta_k^* \downarrow & & \delta_k^* \downarrow \\
H^*(MSp^{\Sigma_k}) & \xrightarrow{(\beta_j^{(k)})^*} & H^*(MSp^{\Sigma_k}) & \xrightarrow{(\beta_j^{(k)})^*} & H^*(MSp^{\Sigma_k}) \\
(\pi_j^{(k)})^* \downarrow & & (\pi_j^{(k)})^* \downarrow & & (\pi_j^{(k)})^* \downarrow \\
H^*(MSp^{\Sigma_{k-1}}) & \xrightarrow{(\beta_j^{(k-1)})^*} & H^*(MSp^{\Sigma_{k-1}}) & \xrightarrow{(\beta_j^{(k-1)})^*} & H^*(MSp^{\Sigma_{k-1}}) \\
\downarrow & & \downarrow & & \downarrow \\
0 & & 0 & & 0
\end{array}
\tag{3.26}
$$

The equality $(\pi_j^{(k)})^*(x_{k-j-1}) = 0$ gives that there exists the element $z \in H^*(MSp^{\Sigma_{k-1}})$ such that $\delta_k^*(z) = x_{k-j-1}$. The exactness of the rows of the diagram (3.27) and induction assumptions imply that there exists the element $y_{k-j-1} \in H^*(MSp^{\Sigma_k})$ such that $(\beta_j^{(k)})^*(y_{k-j-1}) = x_{k-j-1}$. The latter is denoted by $\hat{y} = y_0 + \ldots + y_{k-j-1}$. Then we have

$$
(\beta_j^{(k)})^*(\hat{y}) = (\beta_j^{(k)})^*(y_0) + \ldots + (\beta_j^{(k)})^*(y_{k-j-1})
$$
$$
= (x_0 - x_1) + \ldots + (x_{k-j-2} + x_{k-j-1}) + x_{k-j-1} = x_0.
$$

Lemma 3.4.2. is proved. □

Note also that the Bockstein operators

$$
\beta_j^{(k)} : MSp_*^{\Sigma_k}(\cdot) \longrightarrow MSp_*^{\Sigma_k}(\cdot)
$$

all together induce the transformation

$$\beta^{(k)}(m) : MSp_*^{\Sigma_k(m)}(\cdot) \longrightarrow MSp_*^{\Sigma_k(m)}(\cdot)$$

and the corresponding map of the classifying spectra

$$\beta^{(k)}(m) : MSp^{\Sigma_k(m)} \longrightarrow MSp^{\Sigma_k(m+1)}$$

Lemma 3.4.3 *The sequence*

$$H^*(MSp^{\Sigma_k(1)}) \xleftarrow{\beta^{(k)}(1)} H^*(MSp^{\Sigma_k(2)}) \xleftarrow{\beta^{(k)}(2)} H^*(MSp^{\Sigma_k(3)}) \xleftarrow{\beta^{(k)}(3)} \cdots$$

is a total complex of the lattice of complexes

$$H^*(\mathcal{L}^*_{\Sigma_k}) = \left\{H^*_{\Sigma_k}(\alpha), \beta_1^{(k)}(\alpha), \ldots, \beta_k^{(k)}(\alpha)\right\}_{\alpha\in\mathfrak{A}(k)},$$

where

$$H_*^{\Sigma_k}(\alpha) = H^{-(s-\sum_{i=1}^k a_i(p_i+1))}(MSp^{\Sigma_k})$$

for every collection $\alpha = (a_1, \ldots, a_k) \in \mathfrak{A}(k) \subset \mathfrak{A}$.

Proof. Recall that the lattice $\mathcal{L}^*_{\Sigma_k}$ have been defined in section 1.3. Let us take the following *sublattice* of the lattice $\mathcal{L}_*^{\Sigma_k}$:

$$\widehat{\mathcal{L}}_*^{\Sigma_k}(\cdot) = \left\{\widehat{L}_*^{\Sigma_k}(\sigma)(\cdot), \beta_1^{(k)}(\sigma), \ldots, \beta_k^{(k-1)}(\sigma)\right\}_{\sigma\in\mathfrak{A}(k-1)},$$

where we set

$$\widehat{L}_*^{\Sigma_k}(\sigma)(X,Y) = MSp^{\Sigma_k}_{s-\sum_{i=1}^{k-1} c_i(p_i+1)}(X,Y)$$

for every collection $\sigma = (c_1, \ldots, c_{k-1}) \in \mathfrak{A}(k-1) \subset \mathfrak{A}(k)$. We note that the transformations $\beta_j(\sigma)$ anticommute as in Lemma 1.3.2. Suppose $\widehat{\mathcal{T}}^{\Sigma_k}(\cdot)$ is the total complex of the lattice $\widehat{\mathcal{L}}_*^{\Sigma_k}$:

$$\widehat{MSp}_*^{\Sigma_k(1)}(\cdot) \xrightarrow{d(1)} \widehat{MSp}_*^{\Sigma_k(2)}(\cdot) \longrightarrow \cdots \longrightarrow \widehat{MSp}_*^{\Sigma_k(m)}(\cdot) \xrightarrow{d(m)} \cdots$$

A total complex $\mathcal{T}^{\Sigma_k}(\cdot)$ of the lattice $\mathcal{L}_*^{\Sigma_k}(\cdot)$

$$MSp_*^{\Sigma_k(1)}(\cdot) \xrightarrow{\beta(1)} MSp_*^{\Sigma_k(2)}(\cdot) \longrightarrow \cdots \longrightarrow MSp_*^{\Sigma_k(m)}(\cdot) \xrightarrow{\beta(m)} \cdots$$

coincides with the total complex of the double complex

$$
\begin{array}{ccccc}
\widehat{MSp}_*^{\Sigma_k(1)}(\cdot) & \xrightarrow{d(1)} & \widehat{MSp}_*^{\Sigma_k(2)}(\cdot) \to \cdots \to & \widehat{MSp}_*^{\Sigma_k(m)}(\cdot) & \xrightarrow{d(m)} \cdots \\
\downarrow \widehat{\beta}_k & & \downarrow \widehat{\beta}_k & \downarrow \widehat{\beta}_k & \\
\widehat{MSp}_*^{\Sigma_k(1)}(\cdot) & \xrightarrow{d(1)} & \widehat{MSp}_*^{\Sigma_k(2)}(\cdot) \to \cdots \to & \widehat{MSp}_*^{\Sigma_k(m)}(\cdot) & \xrightarrow{d(m)} \cdots \\
\downarrow \widehat{\beta}_k & & \downarrow \widehat{\beta}_k & \downarrow \widehat{\beta}_k & \\
\vdots & & \vdots & \vdots & \\
\downarrow & & \downarrow & \downarrow & \\
\widehat{MSp}_*^{\Sigma_k(1)}(\cdot) & \xrightarrow{d(1)} & \widehat{MSp}_*^{\Sigma_k(2)}(\cdot) \to \cdots \to & \widehat{MSp}_*^{\Sigma_k(m)}(\cdot) & \xrightarrow{d(m)} \cdots \\
\downarrow \widehat{\beta}_k & & \downarrow \widehat{\beta}_k & \downarrow \widehat{\beta}_k & \\
\vdots & & \vdots & \vdots &
\end{array}
\tag{3.27}
$$

Here the transformations $\widehat{\beta}_k$ are induced by the Bockstein operators β_k (with the corresponding signs; see Lemma 1.3.2):

$$\beta_k^k : MSp_*^{\Sigma_k}(\cdot) \longrightarrow MSp_*^{\Sigma_k}(\cdot).$$

Let $\widehat{MSp}^{\Sigma_k(m)}$ be a classifying spectrum of the bordism theory $MSp_*^{\Sigma_k(m)}(\cdot)$ (which splits into a direct sum of the theories $MSp_*^{\Sigma_k}(\cdot)$)

As a result the complex (3.27) coincides with the total complex of the following double complex:

$$
\begin{array}{ccccccc}
H^*(\widehat{MSp}_*^{\Sigma_k(1)}) & \xleftarrow{d(1)^*} & H^*(\widehat{MSp}_*^{\Sigma_k(2)}) & \leftarrow \cdots \leftarrow & H^*(\widehat{MSp}_*^{\Sigma_k(m)}) & \xleftarrow{d(m)^*} & \cdots \\
\uparrow{\scriptstyle \widehat{\beta}_k^*} & & \uparrow{\scriptstyle \widehat{\beta}_k^*} & & \uparrow{\scriptstyle \widehat{\beta}_k^*} & & \\
H^*(\widehat{MSp}_*^{\Sigma_k(1)}) & \xleftarrow{d(1)^*} & H^*(\widehat{MSp}_*^{\Sigma_k(2)}) & \leftarrow \cdots \leftarrow & H^*(\widehat{MSp}_*^{\Sigma_k(m)}) & \xleftarrow{d(m)^*} & \cdots \\
\uparrow & & \uparrow & & \uparrow & & \\
\vdots & & \vdots & & \vdots & & \\
\uparrow{\scriptstyle \widehat{\beta}_k^*} & & \uparrow{\scriptstyle \widehat{\beta}_k^*} & & \uparrow{\scriptstyle \widehat{\beta}_k^*} & & \\
H^*(\widehat{MSp}_*^{\Sigma_k(1)}) & \xleftarrow{d(1)^*} & H^*(\widehat{MSp}_*^{\Sigma_k(2)}) & \leftarrow \cdots \leftarrow & H^*(\widehat{MSp}_*^{\Sigma_k(m)}) & \xleftarrow{d(m)^*} & \cdots \\
\uparrow & & \uparrow & & \uparrow & & \\
\vdots & & \vdots & & \vdots & &
\end{array}
\tag{3.28}
$$

The rows of the diagram (3.28) are exact due to the induction, the first step of which is provided by Lemma 3.4.2. The latter also implies exactness of the columns. Hence the exactness of the complex (3.28) follows from the spectral sequence for the double complex. □

We note that the homomorphism $\pi(0)^*$ in the diagram

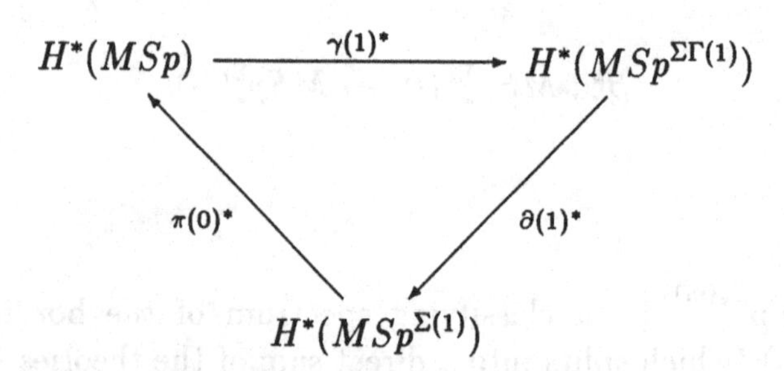

is an epimorphism (as a direct limit of epimorphisms). So $\gamma(1)^* = 0$

and the cohomology groups of the spectrum $MSp^{\Sigma\Gamma(1)}$ are torsion free and are nontrivial only in even dimensions.

Let us have a look at the diagram

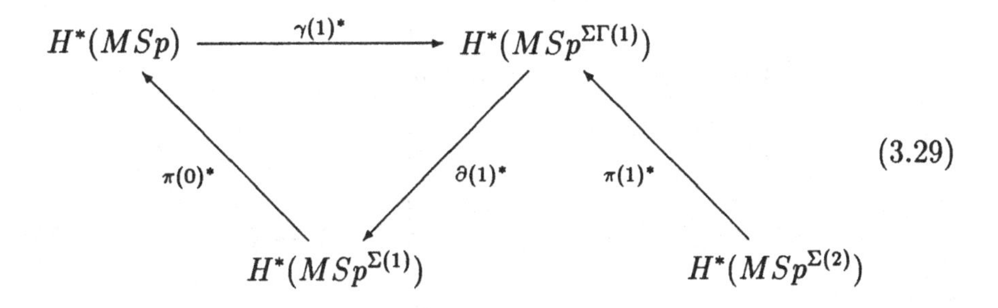

(3.29)

Lemma 3.4.4 *The homomorphism $\pi(1)^*$ is epimorphism.*

Proof. The homomorphism $\partial(1)^*$ is known to be a monomorphism, so it suffices to show that

$$\operatorname{Ker} \pi(1)^* \subseteq \operatorname{Im} \beta(1)^*.$$

Let $x \in H^*(MSp^{\Sigma})$, $\pi(1)^*x = 0$. We take k, such that $x_k = \pi_k^*(x) \neq 0$, $\pi_{k-1}^*(x) = 0$, where the map

$$\pi_k : MSp^{\Sigma_k} \longrightarrow MSp^{\Sigma}$$

is a direct limit:

$$\pi_k = \varinjlim_i (\pi_{k+1}^k \circ \ldots \circ \pi_{k+1+i}^{k+i}).$$

Then we need the following diagram:

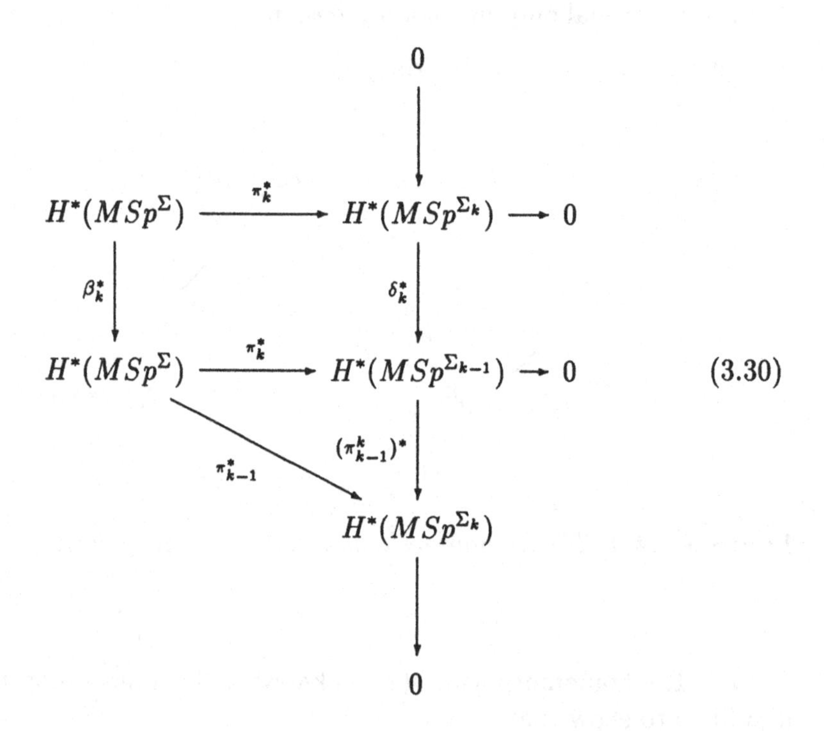

(3.30)

It follows from exactness of the right column that there exists the element $y \in H^*(MSp^\Sigma)$, such that $x - \beta_k^*(y) \in \operatorname{Ker} \pi_k^*$. Let $x_1 = x - \beta_k^*(y)$; then we take $k_1 > k$ such that $\pi_{k_1}^*(x_1) = 0$, $\pi_{k_1-1}^*(x_1) \neq 0$. Repeating the procedure gives that there exists the element $y_1 \in H^*(MSp^\Sigma)$, such that $x_1 - \beta_{k_1}^*(y_1) \in \operatorname{Ker} \pi_{k_1}^*$. We recall that the homomorphism π_n^* is an isomorphism in dimensions up to $2^{n+2} - 3$, so repeating the above procedure gives the elements $x_0 = x, x_1, \ldots, x_l, y_0 = y, y_1, \ldots, y_l$, such that

$$x_i = x_{i-1} - \beta_{k_i-1}^*(y_{i-1}) \in \operatorname{Ker} \pi_{k_i-1}^*, \quad i = 1, \ldots, l,$$

$$\beta_{k_l}^*(y_l) = x_l = x_{l-1} - \beta_{k_l-1}^*(y_{l-1}).$$

The definition of the transformation $\beta(1)$ implies that

$$\beta(1)^*(y_0 \oplus \ldots \oplus y_l) = \sum_{i=0}^{l} \beta_{k_i}(y_i) = x. \ \square$$

Now we need the following commutative diagram:

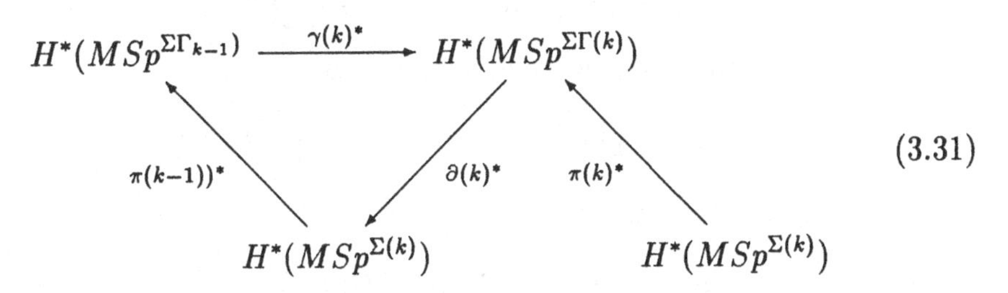

(3.31)

Lemma 3.4.5 *The homomorphism $\pi(k)^*$ is epimorphism.*

Proof. The exactness of the triangle of (3.31) implies that $\partial(k)^*$ is monomorphism and the inclusion Ker $\beta(k-1)^* \subset$ Im $\pi(k-1)^*$. So $\pi(k)^*$ is epimorphism if Im $\beta(k)^* \supset$ Im $\partial(k)^*$. Lemmas 3.4.2, 3.4.3 and exactness of the above triangle imply the equality

$$\text{Im } \beta(k)^* = \text{Ker } \beta(k-1)^* = \text{Ker } \pi(k-1)^* = \text{Im } \partial(k)^*. \ \square$$

In particular, the homomorphisms $\gamma(k)^*$ are proved to be trivial. As noted above, the cohomology groups of the spectra $MSp^{\Sigma\Gamma(k)}$, $MSp^{\Sigma(k)}$ are torsion free and nontrivial only in even dimensions. The Atiyah-Hirzebruch spectral sequence gives the following isomorphisms:

$$MSp^*_\Sigma(MSp^{\Sigma\Gamma(k)}) \cong MSp^*_\Sigma \otimes H^*(MSp^{\Sigma\Gamma(k)}),$$
$$MSp^*_\Sigma(MSp^{\Sigma(k)}) \cong MSp^*_\Sigma \otimes H^*(MSp^{\Sigma(k)}). \qquad (3.32)$$

So the homomorphisms $MSp^*_\Sigma(\gamma(k))$ are trivial and the complex (3.24) is a projective resolution (its exactness follows from Lemma 3.4.2 and isomorphisms (3.33)). Theorem 3.4.1 is proved. □

Corollary 3.4.6 *The Σ-singularities spectral sequence for the symplectic cobordism ring coincides with the Adams-Novikov spectral sequence for the spectrum MSp in the cohomology theory $MSp^*_\Sigma(\cdot)$.* □

The cobordism theory $MSp^*_\Sigma(\cdot)$ possesses some commutative and associative product structure. Therefore the Adams-Novikov spectral sequence also possesses a product structure generated by the *admissible* product structure in the cobordism theory $MSp^*_\Sigma(\cdot)$.

In particular, we have the Adams filtration

$$MSp_* \xleftarrow{\gamma(1)} MSp_*^{\Sigma\Gamma(1)} \longleftarrow \cdots \longleftarrow MSp_*^{\Sigma\Gamma(k)} \xleftarrow{\gamma(k)} \cdots$$

It follows that the isomorphism

$$MSp_*/Tors \cong MSp_*/MSp_*^{\Sigma\Gamma(1)}$$

is naturally implied by the definition of the filtration.

Corollary 3.4.7 *Every element $x \in Tors\, MSp_*$ can be represented by a $\Sigma\Gamma(k)$-manifold for some $k \geq 1$.* □

Note 3.4.1 *By now (March, 1991) all the torsion elements of the ring MSp_* are known to be generated by the Ray elements and Massey products containing them.*

It seems convenient to interpret $\Sigma\Gamma(k)$-manifolds as nonlinear Massey brackets. □

So we have

Corollary 3.4.8 *The torsion $Tors\, MSp_*$ is generated by the elements*

$$\theta_1,\ \varphi_1, \ldots, \varphi_{2^k-2}, \ldots$$

and by the nonlinear Massey brackets containing them. □

3.5 Splitting of the spectrum $MSp^\Sigma_{(2)}$

According to Boardman [13] the spectrum $MSp^\Sigma_{(2)}$ splits into the wedge of suspensions of the spectrum BP. We are to take a particular splitting, since it is necessary to identify the Σ-singularities spectral sequence with the Adams-Novikov one for MSp in the Brown-Peterson cohomology theory $BP^*(\cdot)$.

Let us examine the Hurewicz homomorphism

$$h^{\Sigma} : (MSp^{\Sigma}_{*})_{(2)} \longrightarrow H_{*}(MSp^{\Sigma}_{(2)}; \mathbf{Z})$$

which is clearly monomorphism. So the ring $(MSp^{\Sigma}_{*})_{(2)}$ can be identified with its image in the ring $H_{*}(MSp^{\Sigma}_{(2)}; \mathbf{Z})$.

Theorem 3.5.1 *There exist multiplicative generators*

$$m_i \in H_{2(2^i-1)}(MSp^{\Sigma}_{(2)}; \mathbf{Z})$$

of the polynomial ring

$$H_{*}(MSp^{\Sigma}_{(2)}; \mathbf{Z}) \cong \mathbf{Z}[m_1, \ldots, m_k, \ldots, y_2, \ldots, y_m, \ldots]$$

satisfying Hazewinkel's formula (see [4] *)*

$$v_k = 2m_k - \sum_{i=1}^{k-1} m_i \cdot v_{k-1}^{2^i}, \tag{3.33}$$

where $v_i = \lambda_i w_i + q_i$, $\lambda_i \in \mathbf{Z}_{(2)}$ *are invertible elements, and* $q_i \in (MSp^{\Sigma}_{*})_{(2)}$ *are decomposable ones.*

Proof. Suppose

$$h^{\Sigma} : MSp^{\Sigma}_{(2)} \longrightarrow MSp^{\Sigma}_{(2)} \wedge K(\mathbf{Z})$$

is the spectrum map which generates the Hurewicz homomorphism. We consider the *classic Adams spectral sequences for the spectra* $MSp^{\Sigma}_{(2)}$ and $MSp^{\Sigma}_{(2)} \wedge K(\mathbf{Z})$:

$$Ext^{*,*}_{\mathcal{A}_2}(H^{*}(MSp^{\Sigma}_{(2)}; \mathbf{Z}/2), \mathbf{Z}/2) \Longrightarrow (MSp^{\Sigma}_{*})_{(2)},$$

$$Ext^{*,*}_{\mathcal{A}_2}(H^{*}(MSp^{\Sigma}_{(2)} \wedge K(\mathbf{Z}); \mathbf{Z}/2), \mathbf{Z}/2) \Longrightarrow (MSp^{\Sigma}_{(2)} \wedge K(\mathbf{Z}))_{*}.$$

Theorem 3.3.3 implies the isomorphisms

$$Ext^{*,*}_{\mathcal{A}_2}(H^{*}(MSp^{\Sigma}_{(2)}; \mathbf{Z}/2), \mathbf{Z}/2) \cong Ext^{*,*}_{\mathcal{A}_2}(\mathcal{A}'_2, \mathbf{Z}/2) \otimes \mathbf{Z}[c_2, \ldots, c_n \ldots,],$$

$$\begin{aligned} &Ext^{*,*}_{\mathcal{A}_2}(H^*(MSp^{\Sigma}_{(2)} \wedge K(\mathbf{Z}); \mathbf{Z}/2), \mathbf{Z}/2) \\ &\cong Ext^{*,*}_{\mathcal{A}_2}(\mathcal{A}'_2 \otimes H^*(K(\mathbf{Z}); \mathbf{Z}/2), \mathbf{Z}/2) \otimes \mathbf{Z}[c_2, \ldots, c_n \ldots,], \end{aligned}$$

where $\deg c_n = 4n, n = 2, 4, 5, \ldots, n \neq 2^m - 1$.

We consider the inclusion of $\mathcal{A}'_2$-modules

$$\iota^* : \mathcal{A}'_2 \longrightarrow H^*(MSp^{\Sigma}_{(2)}; \mathbf{Z}/2) \cong \bigoplus_{\omega} \mathcal{A}'_2 \; s_\omega$$

into the direct summand with the generator **1**.

Lemma 3.5.2 *There exists a spectrum map*

$$\sigma : BP \longrightarrow MSp^{\Sigma}_{(2)},$$

such that the $\mathcal{A}'_2$-module homomorphism

$$\mathcal{A}'_2 \xrightarrow{\iota^*} H^*(MSp^{\Sigma}_{(2)}; \mathbf{Z}/2) \xrightarrow{\sigma^*} H^*(BP; \mathbf{Z}/2)$$

is the identity map.

Proof. The map $\sigma : BP \to MSp^{\Sigma}_{(2)}$ determines the element $\tilde{\sigma} \in (MSp^{\Sigma}_{(2)})^*(BP)$.

We note that

$$(MSp^{\Sigma}_{(2)})^*(BP) \cong Hom^*_{A^{BP}}(BP^*(MSp^{\Sigma}_{(2)}), BP^*(BP)). \qquad (3.34)$$

We consider the filtration of the ring BP^* generated by powers of the maximal proper ideal $\mathfrak{m} = (2, v_1, \ldots, v_k, \ldots) \subset BP^*$. As we have seen in section 3.1, this filtration induces the filtration of the algebra (3.35), and the corresponding adjoint object is equal to

$$Hom^*_{\mathcal{A}'_2}(H^*(MSp^{\Sigma}_{(2)}; \mathbf{Z}/2), \mathcal{A}'_2 \otimes \overline{BP^*}), \qquad (3.35)$$

where $\overline{BP}^*$ is the ring adjoined to the ring BP^* with respect to this filtration.

Consider the homomorphism

$$\hat{\sigma} : H^*(MSp^{\Sigma}_{(2)}; \mathbf{Z}/2) \xrightarrow{pr} \mathcal{A}'_2 = \mathcal{A}'_2 \otimes \mathbf{1} \xrightarrow{i} \mathcal{A}'_2 \otimes \overline{BP^*},$$

where *pr* is a projection on the direct summand with the generator **1**, and i is a standard inclusion. The element $\widehat{\sigma}$ of the algebra (3.36) is covered by some element

$$\widetilde{\sigma} \in Hom^{0}_{A^{BP}}(BP^*(MSp^{\Sigma}_{(2)}), BP^*(BP))$$

which determines the element $\widetilde{\sigma} \in (MSp^{\Sigma}_{(2)})^*(BP)$ due to isomorphism (3.35). It is clear that every map of the class $\sigma \in \widetilde{\sigma}$ gives the desired cohomology homomorphism. □

To continue the proof of the theorem we need the commutative diagram

$$\begin{array}{ccc} H^*(MSp^{\Sigma}_{(2)}) & \xrightarrow{h^{\Sigma}} & H^*(MSp^{\Sigma}_{(2)} \wedge K(\mathbf{Z})) \\ \downarrow{\scriptstyle \sigma} & & \downarrow{\scriptstyle \sigma\wedge \mathrm{Id}} \\ BP & \xrightarrow{h^{BP}} & BP \wedge K(\mathbf{Z}) \end{array} \tag{3.36}$$

where σ is the map from Lemma 3.5.2, h^{BP} is the map corresponding to the Hurewicz homomorphism in the Brown-Peterson theory $BP^*(\cdot)$. The homomorphism

$$(h^{BP})^* : H^*(BP; \mathbf{Z}/2) \longrightarrow H^*(BP \wedge K(\mathbf{Z}); \mathbf{Z}/2)$$

was examined in several papers; see, for example, S.Novikov [76]. Suppose

$$h_i \in Ext^{1,2(2^i-1)}_{\mathcal{A}_2}(\mathcal{A}'_2, \mathbf{Z}/2)$$

is a projection of the generators $v_i \in BP_{2(2^i-1)}$, and

$$q_i \in Ext^{1,2(2^i-1)}_{\mathcal{A}_2}(\mathcal{A}'_2 \otimes H^*(K(\mathbf{Z}); \mathbf{Z}/2), \mathbf{Z}/2)$$

is a projection of the standard generators $m_i \in H_*(BP; \mathbf{Z}/2)$ (and q_0 is a projection of $2, \deg q_0 = (0,1)$). According to [76] (see also [103, chapters 4–5]) the induced homomorphism

$$(h^{BP})_* : Ext^{1,2(2^i-1)}_{\mathcal{A}_2}(\mathcal{A}'_2, \mathbf{Z}/2) \longrightarrow Ext^{1,2(2^i-1)}_{\mathcal{A}_2}(\mathcal{A}'_2 \otimes H^*(K(\mathbf{Z}); \mathbf{Z}/2), \mathbf{Z}/2)$$

converts the generators h_i into the elements $q_0 q_i$ for all $i = 1, 2, \ldots$. Commutativity of the diagram (3.36) implies that the homomorphism

$$Ext_{\mathcal{A}_2}^{1,2(2^i-1)}(H^*(MSp^\Sigma; \mathbf{Z}/2), \mathbf{Z}/2)$$

$$\xrightarrow{h_*^\Sigma} Ext_{\mathcal{A}_2}^{1,2(2^i-1)}(\mathcal{A}_2' \otimes H^*(MSp^\Sigma \wedge K(\mathbf{Z}); \mathbf{Z}/2), \mathbf{Z}/2)$$

of the initial terms of the Adams spectral sequences behaves as $(h^{BP})_*$. Recall that the elements $w_i \in MSp_*^\Sigma$ are adjoined to the elements $\sigma_*(h_i)$ by definition (see the proof of Lemma 3.3.3), where

$$\sigma_* : Ext_{\mathcal{A}_2}^{1,2(2^i-1)}(\mathcal{A}_2', \mathbf{Z}/2) \longrightarrow Ext_{\mathcal{A}_2}^{1,2(2^i-1)}(H^*(MSp^\Sigma; \mathbf{Z}/2), \mathbf{Z}/2)$$

is the homomorphism which is induced by the map σ. So the homotopy group homomorphism

$$\sigma_* : BP_* \longrightarrow MSp^\Sigma_{(2)*}$$

converts the elements v_i into the elements w_i up to decomposables. It can be supposed that $\sigma_*(v_i) = \lambda_i w_i + \mathrm{q}_i$, where $\lambda_i \in \mathbf{Z}_{(2)}$ is an invertible scalar, q_i is a decomposable element of the ring $(MSp_*^\Sigma)_{(2)}$. Commutativity of diagram (3.36) allows us to take

$$m_i \in H_{2(2^i-1)}(MSp^\Sigma_{(2)}; \mathbf{Z})$$

such that the Hazewinkel formulas hold. □

Corollary 3.5.3 *There exists a multiplicative projector*

$$\pi^0 : (MSp^\Sigma_{(2)})^*(\cdot) \longrightarrow (MSp^\Sigma_{(2)})^*(\cdot)$$

such that its image is the Brown-Peterson cohomology theory $BP^*(\cdot)$. In addition, the standard Hazewinkel generators v_i of the ring BP^* satisfy the following formula:

$$\pi_*^0(v_i) = \lambda_i w_i + \mathrm{q}_i, \quad i = 1, 2, \ldots,$$

where $\lambda_i \in \mathbf{Z}_{(2)}$ are invertible scalars and q_i are decomposable elements of the ring $(MSp_*^\Sigma)_{(2)}$.

Proof. Let

$$g_{\Sigma}(z) = z + \sum_{i=1}^{\infty} m_i \cdot z^{2^i} \in H^*(MSp^{\Sigma}_{(2)} \wedge K(\mathbf{Z}); \mathbf{Z}/2)\,[[z]]$$

be a formal series, where m_i are the elements of formula (3.34). The series $g_{\Sigma}(z)$ determines the first Chern class c_{Σ} in the cohomology theory $(MSp^{\Sigma}_{(2)})^*(\cdot)$. Suppose $\mathbf{F}_{\Sigma}(x,y) = g_{\Sigma}^{-1}(g_{\Sigma}(x) + g_{\Sigma}(y))$ is a corresponding formal group.

Universality of the first Chern class gives that the class c_{Σ} is determined by some map:

$$\sigma : BP \longrightarrow MSp^{\Sigma}_{(2)}.$$

The standard results of Formal Group Theory give that the coefficients of the formal group $\mathbf{F}_{\Sigma}(x,y)$ generate the ring

$$BP^* = \mathbf{Z}_{(2)}\,[v_1, \ldots, v_k, \ldots]\,,$$

where $v_i = \sigma_*(\lambda_i w_i + q_i)$. Boardman's Theorem B [13] implies that the first Chern class determines the splitting

$$MSp^{\Sigma}_{(2)} = BP \wedge M(G),$$

where $G = \mathbf{Z}_{(2)}\,[x_2, \ldots, x_n, \ldots]$, $n = 2, 4, 5, \ldots, n \neq 2^m - 1, \deg x_n = 4n$, $M(G)$ is a graded Moore space. □

Note 3.5.1 *So the theory $BP^*(\cdot)$ is represented as a direct summand of the cobordism theory $(MSp^{\Sigma}_{(2)})^*(\cdot)$. It would seem that we can find a certain splitting of the theory $(MSp^{\Sigma}_{(2)})^*(\cdot)$ into the sum of the theories $BP^*(\cdot)$ such that the generators w_i would be exactly covered by the elements v_i of the ring BP_*. However we cannot; see V.Gorbunov* [42]. *His proof is essentially based on a description of the Ray elements in terms of the Two-valued Formal Group Theory (V.Buchstaber* [26, Theorem 23.11]*).* □

According to Theorem 3.1.3 we have

Corollary 3.5.4 *A 2-localization of the diagram* (3.22) *is an Adams resolution of the spectrum MSp in the theory $BP^*(\cdot)$, and a localized Σ-singularities spectral sequence for the symplectic cobordism ring coincides with the Adams-Novikov spectral sequence for the spectrum MSp in the theory $BP^*(\cdot)$.* □

A few remarks concerning Σ-singularities spectral sequence would come in handy.

Let $u_1, u_2, \ldots, u_k, \ldots$ be the projections of the elements $\theta_1, \varphi_1, \ldots, \varphi_{2^k}, \ldots$ into the first term $E_1^{*,*}$ of our spectral sequence. As in section 1.6, we have the isomorphism

$$E_1^{*,*} = (MSp_*^{\Sigma})_{(2)}\left[u_1, u_2, \ldots, u_k, \ldots\right].$$

The line $E_1^{s,*} = (MSp_*^{\Sigma(s+1)})_{(2)}$ is a free module over the ring $(MSp_*^{\Sigma})_{(2)}$ with the generators

$$u_1^{a_1}, u_2^{a_2}, \ldots, u_k^{a_k}, \ldots,$$

where $a_1 \geq 0, a_1 + \ldots + a_k + \ldots = s + 1$. Corollary 3.4.5 implies that the algebra

$$Ext_{A^{BP}}^{*,*}(BP^*(MSp_{(2)}^{\Sigma}), BP^*)$$

coincides with the homology algebra of the complex

$$(MSp_*^{\Sigma(1)})_{(2)} \xrightarrow{\beta(1)} (MSp_*^{\Sigma(2)})_{(2)} \longrightarrow \cdots \longrightarrow (MSp_*^{\Sigma(2)})_{(k)} \xrightarrow{\beta(k)} \cdots \tag{3.37}$$

Recall also that the differentials $\beta(k)$ are determined by the Bockstein operators β_i as follows:

$$\beta(k)\left(x \cdot u_1^{a_1}, u_2^{a_2}, \ldots, u_k^{a_k}, \ldots\right) \tag{3.38}$$

$$= \sum_{i=1}^{\infty}\left((-1)^{\epsilon_i(\alpha)}\beta_i x \cdot \left(u_1^{a_1}\ldots u_{i-1}^{a_{i-1}} u_i^{a_i+1} u_{i+1}^{a_{i+1}} \ldots u_k^{a_k}\ldots\right)\right),$$

where $\alpha = (a_1, \ldots, a_i, \ldots)$ and the sign $\epsilon_i(\alpha)$ was determined in section 1.3.

Now we are able to translate the above results concerning the modified algebraic spectral sequence into geometric language. The projector

π^0 from Corollary 3.5.3 actually converts *maSS*-filtration of the ring BP_* into some filtration of the ring $(MSp^{\Sigma}_*)_{(2)}$ (which is also called *maSS*-filtration). Recall that $\pi^0_*(v_i) = \lambda_i w_i + \mathrm{q}_i$, where $\lambda_i \in \mathbf{Z}_{(2)}$ are invertible elements and q_i are decomposable ones of the ring $(MSp^{\Sigma}_*)_{(2)}$. So the *maSS*-filtration coincides with the filtration determined by the graduation ω^{Σ} for the generators of the ring $(MSp^{\Sigma}_*)_{(2)}$:

$$\omega^{\Sigma}(2) = 2, \quad \omega^{\Sigma}(w_i) = 1, \quad i = 1, 2, \ldots, \omega^{\Sigma}(x_n) = 0, \quad n \neq 2^m - 1.$$

Let $(\overline{MSp^{\Sigma}_*})_{(2)}$ be the ring which is associated to the ring $(MSp^{\Sigma}_*)_{(2)}$ with respect to *maSS*-filtration. We have the isomorphism

$$(\overline{MSp^{\Sigma}_*})_{(2)} \cong (\mathbf{Z}/2)\,[h_0, h_1, \ldots, h_k, \ldots, c_2, \ldots, c_n, \ldots],$$

where h_0 is a class of two, h_k is a class containing the element w_k, $k = 1, 2, \ldots$ and c_n is a class of the element x_n, $n = 2, 4, 5, \ldots$, $n \neq 2^m - 1$. Suppose $\mathcal{U}_1, \ldots, \mathcal{U}_n, \ldots$ are a projection of the elements $u_1, \ldots, u_n, \ldots$ into the term $\mathsf{E}_1^{*,*,*}$ of *maSS*. So we get the isomorphism

$$\mathsf{E}_1^{*,*,*} \cong (\overline{MSp^{\Sigma}_*})_{(2)}\,[\,\mathcal{U}_1, \ldots, \mathcal{U}, \ldots].$$

Suppose $\overline{\beta}_k$ is the operator which is associated to the Bockstein operator β_k with respect to *maSS*-filtration for every $k = 1, 2, \ldots$. Then the complex

$$(\overline{MSp^{\Sigma(1)}_*})_{(2)} \xrightarrow{\overline{\beta(1)}} (\overline{MSp^{\Sigma(2)}_*})_{(2)} \longrightarrow \cdots \longrightarrow (\overline{MSp^{\Sigma(k)}_*})_{(2)} \xrightarrow{\overline{\beta(k)}} \cdots$$

associated to the complex (3.38) has the differentials $\overline{\beta(k)}$ acting as follows:

$$\overline{\beta(k)}\,(x \cdot \mathcal{U}_1^{a_1}, \mathcal{U}_2^{a_2}, \ldots, \mathcal{U}_k^{a_k}, \ldots) \tag{3.39}$$
$$= \sum_{i=1}^{\infty} (-1)^{\epsilon_i(\alpha)} \beta_i x \cdot \left(\mathcal{U}_1^{a_1} \ldots \mathcal{U}_{i-1}^{a_{i-1}} \mathcal{U}_i^{a_i+1} \mathcal{U}_{i+1}^{a_{i+1}} \ldots \mathcal{U}_k^{a_k} \ldots\right).$$

Now it is very useful to compare (3.40) with the formula for the first differential $d_1^{a l}$ of the *maSS* of Theorem 3.2.2. We note that

$$\beta_i w_i = 2, \qquad \beta_j w_i = 0, \quad i \neq j,$$

by definition of the elements $w_i \in MSp_*^\Sigma$.

So it is not surprising that $\overline{\beta}_i h_i = h_0$, $\overline{\beta}_j h_i = 0$ for $i \neq j$. Here we use the notation of section 3.2 to formulate

Corollary 3.5.5 *There exist the polynomial generators c_n of the ring*

$$(\overline{MSp_*^{\Sigma(1)}})_{(2)} \cong (\mathbf{Z}/2)\,[h_0, h_1, \ldots, h_k, \ldots, c_2, \ldots, c_n, \ldots]$$

such that the operators $\overline{\beta}_k$ act on them as follows:

1. *if $n = 2^{i-1} + 2^{j-1} - 1$, then*

$$\overline{\beta}_k(c_{i,j}) = \left\{ \begin{array}{ll} h_i & \text{if } k = j, \\ h_j & \text{if } k = i, \\ 0 & \text{if } k \neq i, j; \end{array} \right\} \tag{3.40}$$

2. *if $n = 2^{i_1-1} + \ldots + 2^{i_q-1} - 1$, $q \geq 3$, then*

$$\overline{\beta}_k(c_{i_1,\ldots,i_q}) = \left\{ \begin{array}{ll} h_1 c_{i_1,\ldots,\widehat{i_t},\ldots,i_q} & \text{if } k = i_t, \\ \sum_{t=1}^{q} h_{i_t} c_{i_1,\ldots,\widehat{i_t},\ldots,i_q} & \text{if } k = 1, \\ 0 & \text{if } k \neq 1, i_1, \ldots, i_q; \end{array} \right\} \tag{3.41}$$

3. *if n is even and isn't a power of two then $\overline{\beta}_k(c_n) = 0$ for all $k = 1, 2, \ldots$.* □

Note 3.5.2 *The formula for the first differential of the maSS partly describes the action of the Bockstein operators β_k on the generators of the ring MSp_*^Σ. It is the starting point for correcting the admissible product structure in the cobordism theory $MSp_*^\Sigma(\cdot)$ and for choosing generators of the ring MSp_*^Σ in Chapter 4. As a result the exact formulas for the action of the Bockstein operator β_k will be obtained. This means that we'll have a complete description of the first differential in the Adams-Novikov spectral sequence.* □

Chapter 4

First differential of ANSS

The purpose of this chapter is to combine the geometric machinery of the Σ-*SSS* with the algebra of the *ANSS* for the spectrum MSp.

The first stage is to choose an admissible product structure in the theory $MSp_*^{\Sigma}(\cdot)$, such that the multiplication formulas for the Bockstein operators have the simplest form. We emphasize that the choice will be based on specific properties of symplectic cobordism theory, and in particular depends essentially on known information concerning Ray elements. To achieve the desired product structure we take a relevant algebraic result of Two-valued Formal Group Theory.

Then a localization procedure is applied to compute the action of the Bockstein operators β_k on the generators of the ring MSp_*^{Σ} (Theorem 4.4.1). The computations of the *maSS* are also used in an essential way.

Finally we get a complete description of the first differential of the *ANSS*. At the end we discuss the simplest application of the results achieved of the algebra

$$Ext_{\mathbf{A}^{BP}}^{*,*}(BP^*(MSp^{\theta_1}), BP^*).$$

The main conclusion we come to is that the algebra possesses a module structure over the symmetric group S_∞, hence it may be completely described in terms of the representation theory of the symmetric group.

4.1 Characterization of the Ray elements

As noted above the Ray elements are very important for a description of the *ANSS*. We begin by recalling some definitions.

Let $M(U,Sp)_*(\cdot)$ be the bordism theory of U-manifolds with fixed Sp-structure on the boundary. We have the following exact sequence:

$$0 \longrightarrow MU_{4k-2} \xrightarrow{\rho_*} M(U,Sp)_{4k-2} \xrightarrow{d_*} M(U,Sp)_{4k-2} \longrightarrow 0. \qquad (4.1)$$

J.Alexander [6] showed that the elements

$$\nu_1 \in M(U,Sp)_2, \quad \nu_{2k} \in M(U,Sp)_{8k-3},$$

such that $d_*(\nu_1) = \theta_1$, $d_*(\nu_{2k}) = \varphi_k$, have infinite order. The equalities $2\theta_1 = 0$, $2\varphi_k = 0$ imply that there exist the elements δ_1, δ_{2k} of the group MU_*, such that $\rho_*(\delta_1) = 2\nu_1$, $\rho_*(\delta_{2k}) = 2\nu_{2k}$. It is clear that the elements δ_1, δ_{2k} are uniquely determined in the group $MU_* \otimes (\mathbf{Z}/2)$.

Note 4.1.1 *The elements θ_{2k+1} are trivial for every $k \geq 1$ (Roush [91]). So the elements $\nu_{2k+1}, \delta_{2k+1}$ are assumed to be zero.* □

The exact sequence (4.1) induces the following exact triangle of the classifying spectra:

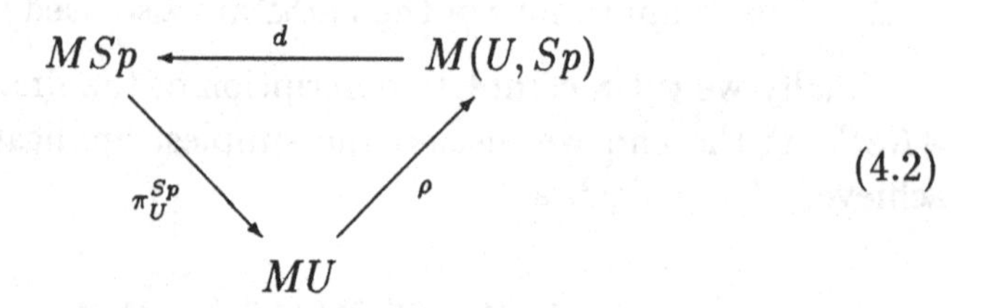

(4.2)

where π_U^{Sp} is a standard map of the classifying spectra. The Atiyah-Hirzebruch spectral sequence gives $MU^*(d) = 0$. So the triangle (4.2) may be considered as the first stage in setting up the Adams resolution of the spectrum MSp in the theory $MU^*(\cdot)$. The triangle (4.2) may be extended up to a certain Adams resolution:

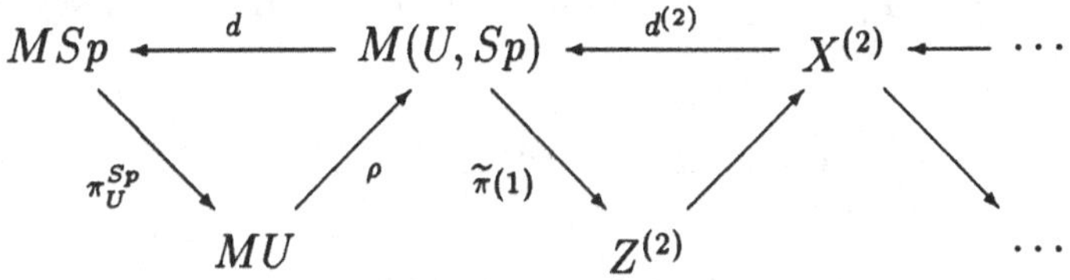

$$\begin{array}{ccccccc} MSp & \xleftarrow{d} & M(U,Sp) & \xleftarrow{d^{(2)}} & X^{(2)} & \longleftarrow & \cdots \\ & \searrow^{\pi_U^{Sp}} \quad \nearrow^{\rho} & & \searrow^{\widetilde{\pi}(1)} \quad \nearrow & & \searrow & \\ & MU & & Z^{(2)} & & & \cdots \end{array} \tag{4.3}$$

The part of the above diagram,

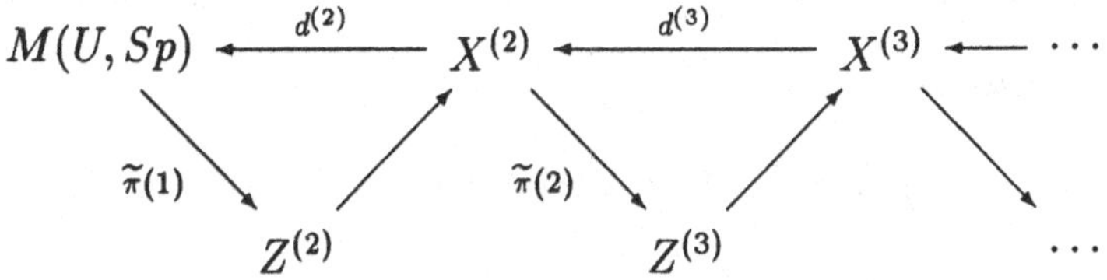

$$\begin{array}{ccccccc} M(U,Sp) & \xleftarrow{d^{(2)}} & X^{(2)} & \xleftarrow{d^{(3)}} & X^{(3)} & \longleftarrow & \cdots \\ & \searrow^{\widetilde{\pi}(1)} \quad \nearrow & & \searrow^{\widetilde{\pi}(2)} \quad \nearrow & & \searrow & \\ & Z^{(2)} & & Z^{(3)} & & & \cdots \end{array} \tag{4.4}$$

is also the Adams resolution of the spectrum $M(U,Sp)$ in the theory $MU^*(\cdot)$. The elements ν_1, ν_{2k} have infinite order, so their images in the group $\pi_* Z^{(2)}$ are not trivial. By definition the elements $\theta_1 = \widetilde{\pi}(1)(\nu_1)$, $\varphi_k = \widetilde{\pi}(1)(\nu_{2k})$ coincide with a projection of the Ray elements into the first line $E_1^{1,*}$ of the initial term of the *ANSS*. The first differential acts on these elements as follows:

$$d_1(\delta_i) = (\widetilde{\pi}(1)_* \circ \rho_*)(\delta_i) = \left\{ \begin{array}{ll} 2\theta_1 & \text{if } i=1, \\ 2\phi_k & \text{if } i=2k. \end{array} \right\} \tag{4.5}$$

So the elements δ_1, δ_{2k} are of order two in the line $E_2^{1,*}$. The elements δ_1, δ_{2k} were computed by V.Buchstaber in terms of Two-valued Formal Group Theory. Now we recall some of his results.

Let $F(u_1,u_2)$ be a formal group in complex cobordism theory, and $\overline{u} \in MU_*[[u]]$ be a series which is the solution of the equation $F(u,\overline{u}) = 0$. Then we have

$$G(x) = u + \overline{u} \in MU_*[[x]],$$

where $x = u\overline{u}$. Let's take the formal series

$$X^+ = F(u_1,u_2)\cdot F(\overline{u}_1,\overline{u}_2), \quad X^- = F(u_1,\overline{u}_2)\cdot F(\overline{u}_1,u_2),$$

and introduce the following notation:

$$\Theta_1(x_1,x_2) = X^+ + X^-, \quad \Theta_2(x_1,x_2) = X^+ \cdot X^-,$$

where $x_i = u_i \cdot \overline{u}_i$, $i = 1, 2$, and the series $\Theta_1(x_1, x_2)$, $\Theta(x_1, x_2)$ belong to the ring $MU_*[[x_1, x_2]]$; see [30]. These series determine the two-valued formal group:

$$\mathfrak{F}^2 - \Theta_1(x_1, x_2)\mathfrak{F} + \Theta_2(x_1, x_2) = 0.$$

The series $\Theta_1(x_1, x_2)$, $G(x)$ are related as follows:

$$\Theta_1(x_1, x_2) = 2A_0(x_1, x_2) + A_1(x_1, x_2)G(x_1)G(x_2). \tag{4.6}$$

It is convenient to set

$$A(x) = A_1(x, 0), \quad \delta(x) = G(x)A(x).$$

Theorem 4.1.1 (V.Buchstaber [26], [27]) *The coefficients of the series $\delta(x) \in MU_*[[x]]$ coincide with the elements δ_i modulo two for all $i = 1, 2, \ldots$.* □

Note 4.1.2 *We know that $\delta_{2i+1} = 0$ for all $i = 1, 2, \ldots$. So we can identify the elements $\delta_k \in MU_*$ with the corresponding coefficients of the series $\delta(x)$.* □

Note 4.1.3 *Let us take Sp-manifolds N_k, such that $[N_1] = \theta_1$, $[N_k] = \varphi_{k-1}$, $k = 2, 3, \ldots$, and Sp-manifolds V_k which bounds two copies of the manifold N_k. The Sp-manifolds V_k may be chosen arbitrarily. Indeed let V_k' be different manifolds, such that $\partial V_k' = 2N_k$, $k = 1, 2, \ldots$. The exact sequence* (4.1) *implies that there exist (U, Sp)-manifolds D_k' bounding the manifolds N_k, such that bordism classes of the manifolds*

$$[D_k' \cup -V_k' \cup D_k']_{MU}$$

coincide with the elements δ_k in the group MU_* for every $k \geq 1$. □

Following V.Buchstaber we denote the ring

$$E_2^{0,*} = Hom^*_{A^{MU}}(MU^*(MSp), MU^*) \subset MU_*$$

by $Л_*$ to formulate the following technical lemma.

Lemma 4.1.2 *There exist elements a_i in the ring MU_* such that the elements $\delta_i^2 + 4a_i$ belong to the ring $Л_*$ for every $i = 1, 2, \ldots$.* □

Proof. The series $\Theta_1(x_1, x_2)$, $\Theta_1(x_1, x_2)$ have the following form:

$$\Theta_1(x_1, x_2) = 2x_1 + 2x_2 + \sum_{i+j \geq 1} \alpha_{i,j} x_1^i x_2^j,$$

$$\Theta_2(x_1, x_2) = (x_1 - x_2)^2 + \sum_{i+j > 1} \beta_{i,j} x_1^i x_2^j.$$

The *discriminant* of the two-valued formal group in the complex cobordism theory has the form

$$\Theta_1^2(x_1, x_2) - 4\Theta_2(x_1, x_2) = H(x_1, x_2) \cdot \mathfrak{q}(x_1) \cdot \mathfrak{q}(x_2), \qquad (4.7)$$

where

$$H(x_1, x_2) = \frac{A_1(x_1, x_2)}{A(x_1) \cdot A(x_1)}, \quad \mathfrak{q}(x) = \delta^2(x) - 4A(x).$$

By differentiating (4.7) with respect to x_1 and supposing $x_2 = 0$ we get

$$\mathfrak{q}(x) = -4x + \sum_{i \geq 2} (\beta_{1,i} - \alpha_{1,i-1}) x^i. \qquad (4.8)$$

The equality (4.8) implies

$$\left(\sum_{i=1}^{\infty} \delta_i x^i \right)^2 - 4x \cdot A(x) \in Л_* [[x]], \qquad (4.9)$$

because the coefficients of the series $\Theta_1(x_1, x_2)$, $\Theta_2(x_1, x_2)$ belong to the ring $Л_*$. Setting

$$x \cdot A(x) = \sum_{i \geq 2} c_i x^i$$

we rewrite the inclusion (4.9) as follows:

$$\delta_i^2 + 2 \sum_{j=1}^{i-1} \delta_{j-i} \delta_i + 4c_i \in Л_*. \qquad (4.10)$$

Formula (4.6)) gives the equality

$$\frac{\Theta_1(x_1,x_2)}{H(x_1,x_2)} = \delta(x_1)\cdot\delta(x_2) + 2C(x_1,x_2).$$

The coefficients of the series $\Theta_1(x_1,x_2)$, $H(x_1,x_2)$ belong to the ring $Л_*$, so there exist elements $b_{i,j} \in MU_*$ such that the following inclusion holds:

$$\delta_j\delta_i + 2b_{i,j} \in Л_*. \tag{4.11}$$

The inclusions (4.10), (4.11) imply that there exist elements $a_i \in Л_*$, such that $\delta_i^2 + 4a_i \in Л_*$. □

4.2 Product structure in $MSp_*^\Sigma(\cdot)$

Recall (Chapter 2) that the admissible product structure in the theory $MG_*^\Sigma(\cdot)$ depends on the choice of the manifolds Q_k, bounding the obstructions P_k', $k = 1, 2, \ldots$. It is clear that the existence of manifolds Q_k with *a minimal number of singularities* is determined by specific properties of the cobordism theory $MG_*(\cdot)$ and the manifolds P_k. Such manifolds Q_k are assumed to exist in the theory $MSp_*^\Sigma(\cdot)$. To prove this we shall resort to identifying the *ANSS* and the Σ-*SSS* for the symplectic cobordism ring, and make use of section 4.1 as well as the following geometric consideration.

Let N_k be Sp-manifolds such that $[N_1] = \theta_1$, $[N_k] = \varphi_{k-1}$ when $k = 2, 3, \ldots$, where φ_j are the Ray elements. See [88], [90] for their description. Let us take Sp-manifolds V_i, such that $\partial V_i = N_i^{(1)} \cup N_i^{(2)}$ where $N_i^{(t)}$ are copies of the manifold N_i, $t = 1, 2$. There are the following Sp-manifolds:

$$R_i = (V_i \times N_i) \,/\, \left(N_i^{(1)} \times N_i \cup N_i \times N_i^{(2)}\right),$$

for every $i = 1, 2, \ldots$. Next we take (U, Sp)-manifolds D_i, bounding the manifolds $N_i : \partial D_i = N_i$, $i = 1, 2, \ldots$. Also we take the following U-manifold:

$$M_i = D_i^{(1)} \cup -V_i \cup D_i^{(1)}.$$

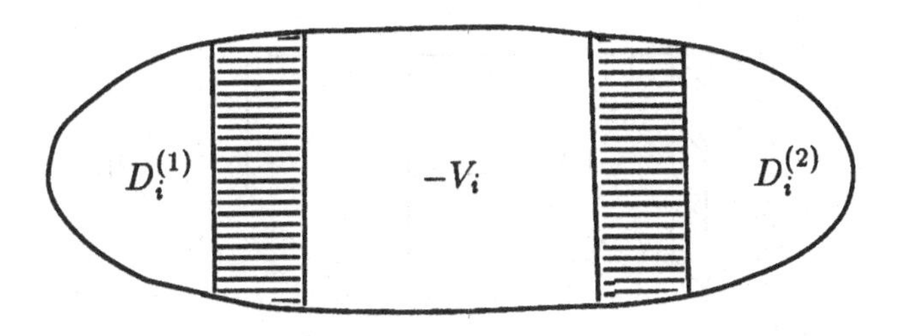

Figure 4.1: Manifold M_i.

All the above manifolds are supposed to have sufficiently large collars; see Figure 4.1. Here $\partial D_i^{(1)} = N_i^{(1)}$, $\partial D_i^{(2)} = N_i^{(2)}$. Now we need some (U, Sp)-manifold bounding the manifold R_i. We set

$$U_i = D_i^{(1)} \times D_i^{(2)} \cup -V_i \cup D_i,$$

where we identify the following submanifolds of the boundaries:

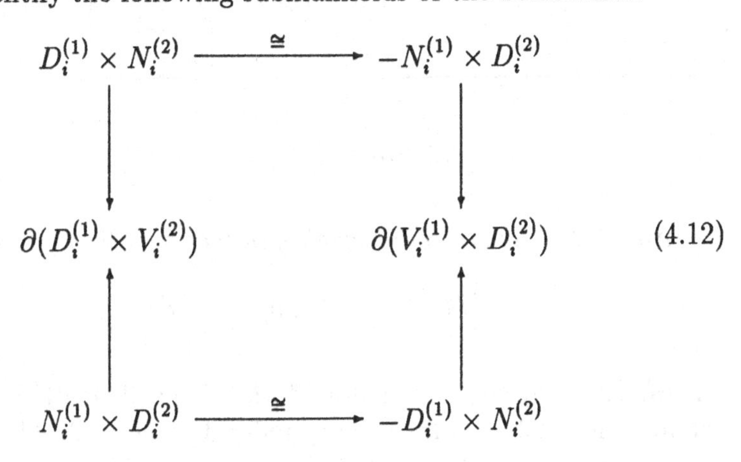

(4.12)

Thus $\partial U_i = R_i$. Finally we take the (U, Sp)-manifold

$$T_i = D_i \times N_i \times S^1,$$

where the Sp-structure on the circle S^1 is such that $[S^1]_{Sp} = \theta_1$, i.e.

$$[\partial T_i]_{Sp} = \varphi_{i-1}^2 \theta_1, i \geq 2.$$

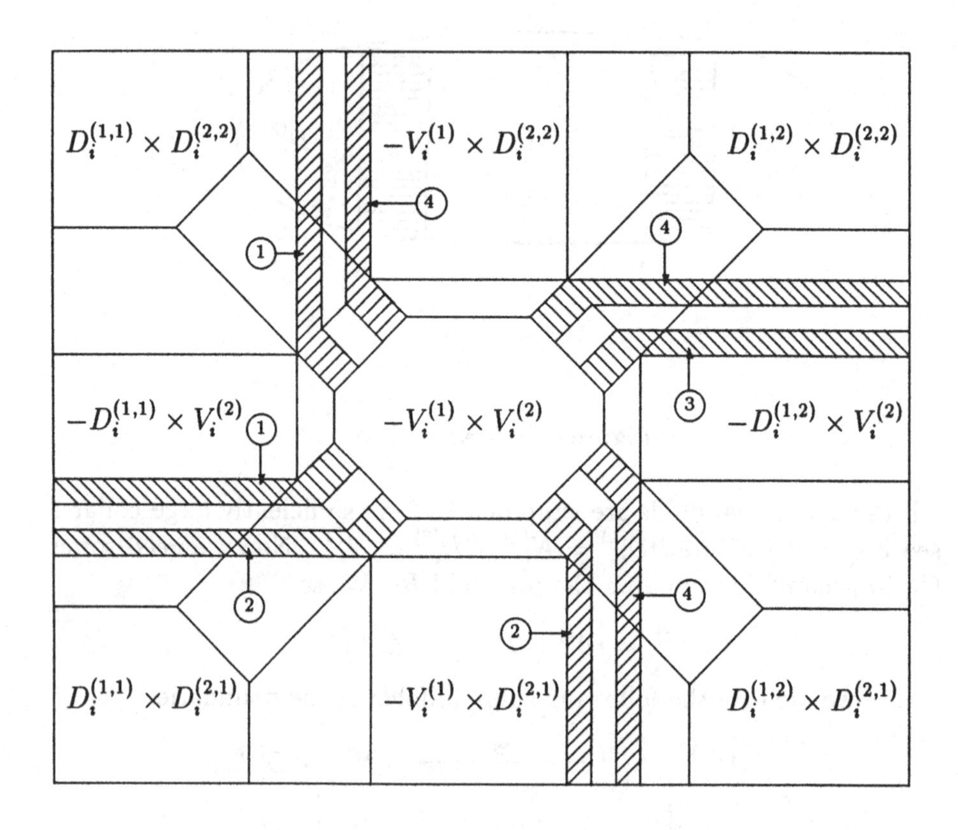

Figure 4.2: $M_i \times M_i \times \{1\}$

Lemma 4.2.1 *There is the equality in the cobordism group $M(U, Sp)_*$,*

$$\left[M_i^2\right]_{U,Sp} = 4\,[U_i]_{U,Sp} + [T_i]_{U,Sp}\,.$$

Proof. Knowing that the manifold M_i has a sufficiently large collar we examine the bottom side of the cylinder $M_i \times M_i \times I$. This manifold is glued together as in Figure 4.2.

The manifold corresponding to the central part of the figure is an Sp-manifold. The manifolds corresponding to its numbered parts are diffeomorphic to the manifold

$$N_i^{(1)} \times N_i^{(2)} \times I_1 \times I_2,$$

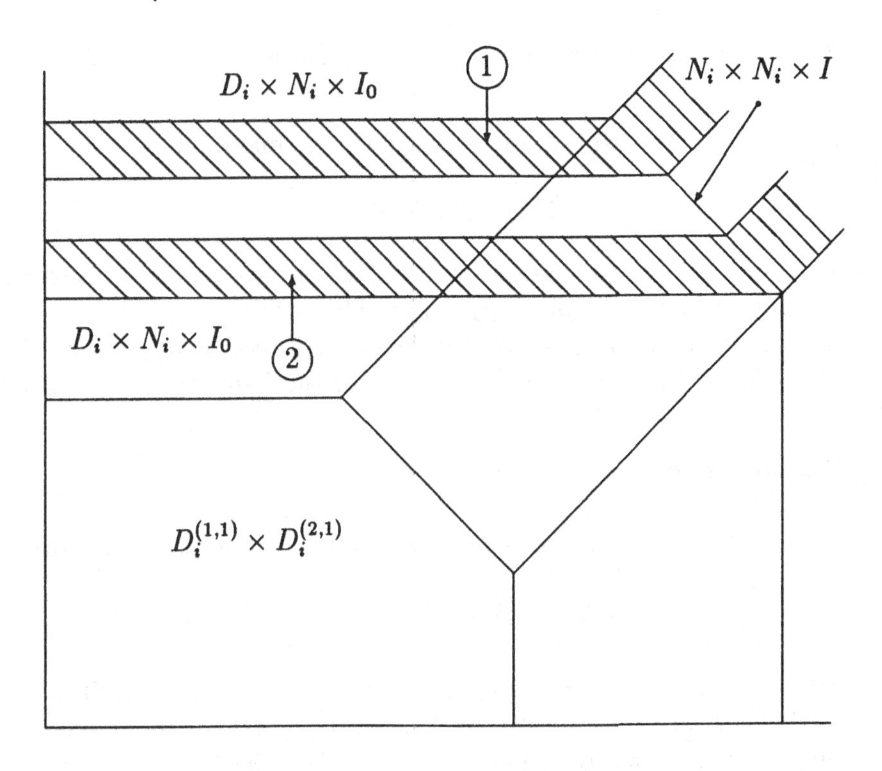

Figure 4.3: A corner of $M_i \times M_i \times \{1\}$.

preserving Sp-structure.

Now we glue together the manifolds having the same numbers by means of the following diffeomorphism:

$$(n_1, n_2, t_1, t_2) \longmapsto (n_2, n_1, t_1, t_2),$$

where $n_s \in N_i^{(s)}$, $t_s \in I_s$, $s = 1, 2$. This diffeomorhism preserves the Sp-structure as well. The corner part of Figure 4.2 is shown in Figure 4.3. The manifolds corresponding to the hatched strips are diffeomorphic to the manifold $D_i \times N_i \times I_0$.

As a consequence every boundary $N_i \times N_i \times I_0$ of the manifold

$$D_i \times N_i \times I_0$$

is glued together with the manifold $N_i \times N_i \times I_0$. So it is possible to identify the manifolds corresponding to the identically numbered strips; see Figure 4.2.

In particular the gluing procedure gives the manifold

$$T_i = D_i \times N_i \times S^1,$$

where the circle S^1 has the desired frame as a result of 4-fold inverting of the factors in the product $N_i \times N_i$. The cylinder $M_i \times M_i \times I$ is the required manifold which establishes (U, Sp)-bordism between the manifold $M_i \times M_i$ and the disjoint union of the U-manifolds $4U_i \cup T_i$. □

It is an easy exercise to prove

Lemma 4.2.2 *There is the equality in the bordism group $M(U, Sp)_*$,*

$$2\,[T_i]_{U,Sp} = 0, \qquad i = 1, 2, \ldots\ . \ \square$$

Let $\alpha_i = [R_i]$ be the corresponding elements of the ring MSp_*. For dimensional reasons $\alpha_i \in Tors\ MSp_*$, so the element α_i belongs to the image of the homomorphism d_* of the Adams filtration for every $i = 1, 2, \ldots$:

$$MSp_* \xleftarrow{d_*} M(U, Sp)_* \xleftarrow{d(2)_*} X_*^{(2)} \xleftarrow{d(3)_*} X_*^{(3)} \longleftarrow \cdots,$$

which is induced by the diagram (4.3). According to the definition of the manifolds U_i the elements $u_i = [U_i]_{U,Sp}$ of the group $M(U, Sp)_*$ are such that $d_* u_i = \alpha_i$.

Lemma 4.2.3 *The projection of the elements α_i into the first line*

$$Ext^{1,*}_{A^{MU}}(MU^*(MSp), MU^*)$$

of the second term $E_2^{*,*}$ of the Adams-Novikov spectral sequence is equal to zero for all $i = 1, 2, \ldots$.

Proof. We have the two following possibilities for the element $u_i \in M(U, Sp)_*$:

1° *the element u_i is equal to zero or has finite order;*

2° *the element u_i isn't zero and has infinite order.*

To see what happens in the *first case* we consider the Adams filtration induced by the diagram (4.4):

$$M(U,Sp)_* \xleftarrow{d(2)_*} X_*^{(2)} \xleftarrow{d(3)_*} X_*^{(3)} \xleftarrow{d(4)_*} X_*^{(4)} \longleftarrow \cdots$$

If the element $u_i \in M(U,Sp)_*$ has finite order then

$$u_i \in \mathrm{Im}\left(d_*^{(2)} \longrightarrow M(U,Sp)_*\right).$$

The diagram

$$\begin{array}{ccccc} MSp_* & \xleftarrow{d_*} & M(U,Sp)_* & \xleftarrow{d_*^{(2)}} & X_*^{(2)} \\ & {\scriptstyle (\pi_U^{Sp})_*}\searrow & \nearrow{\scriptstyle \rho_*} \quad {\scriptstyle \widetilde{\pi}(1)_*}\searrow & \nearrow & \\ & MU_* & & Z_*^{(2)} & \end{array} \tag{4.13}$$

gives that $\widetilde{\pi}(1)_*(u_i) = 0$, i.e. the projection of the element α_i into the first line $E_1^{1,*}$ of the *ANSS* is trivial.

In the *second case* the image of the element u_i with respect to the homomorphism

$$M(U,Sp)_* \longrightarrow Hom^*_{A^{MU}}(MU^*(M(U,Sp)), MU^*)$$

is not trivial. For simplicity we will denote this image by the same u_i.

We note that the projection of the element $[T_i]_{M(U,Sp)} \in M(U,Sp)_*$ into the ring

$$Hom^*_{A^{MU}}(MU^*(M(U,Sp)), MU^*)$$

is zero since $2\,[T]_{U,Sp} = 0$ due to Lemma 4.2.2. The cofibration

$$MSp \xrightarrow{\pi_{Sp}^U} MU \xrightarrow{\rho} M(U,Sp)$$

induces the exact sequence

$$0 \longrightarrow Л_* \xrightarrow{(\pi_{Sp}^U)_*} MU_* \xrightarrow{\widetilde{\rho}_*} Hom^*_{A^{MU}}(MU^*(M(U,Sp)), MU^*) \longrightarrow \cdots \tag{4.14}$$

Lemma 2.2.1 gives the equality $\widetilde{\rho}_*(\delta_i^2) = 4u_i$. On the other hand, there is the element $a_i \in MU_*$, such that $\delta_i^2 + 4a_i \in Л_*$ (see Lemma 4.1.2). Exactness of the sequence (4.14) gives $\widetilde{\rho}_*(-a_i) = u_i$. Therefore the first Adams differential kills the projection $\widetilde{\pi}(1)_*(u_i)$ of the element α_i into the line $E_1^{1,*}$ (see the diagram (4.13)):

$$d_1^{Ad}(-a_i) = \widetilde{\pi}(1)_* \circ \widetilde{\rho}_*(-a_i) = \widetilde{\pi}(1)_*(u_i).$$

So the image of α_i is always zero in the line

$$Ext^{1,*}_{A^{MU}}(MU^*(MSp), MU^*). \ \square$$

Now we can come back to the theory $MSp_*^\Sigma(\cdot)$ (see section 3.3). Here we intend to make some extensions to Vershinin's Theorem 3.3.5.

Theorem 4.2.4 *There exists an admissible product structure μ in the symplectic cobordism theory with singularities $MSp_*^\Sigma(\cdot)$ such that*

1° *the coefficient ring MSp_*^Σ is isomorphic to the polynomial ring*

$$\mathbf{Z}\left[w_1, \ldots, w_k, \ldots, x_2, \ldots, x_n, \ldots\right],$$

where the generators w_k may be presented by any Sp-manifolds W_k, such that $\partial W_k = 2P_k$, $\deg w_k = 2(2^k - 1)$, $\deg x_n = 4n$, $k = 1, 2, \ldots$, $n = 2, 4, 5, \ldots$, $n \neq 2^m - 1$,

2° *for every $k = 1, 2, \ldots$ the Bockstein operator β_k satisfies the product formula*

$$\beta_k(x \cdot y) = \beta_k(x) \cdot y + x \cdot \beta_k(y) - w_k \cdot \beta_k(x)\beta_k(y), \tag{4.15}$$

where $x, y \in MSp_*^\Sigma$.

Proof. In terms of the sequence $\Sigma = (P_1, \ldots, P_k, \ldots)$ we have

$$P_1 = N_1, \quad P_k = N_{2^k - 2}, \quad k \geq 2.$$

Denoting

$$W_1 = V_1, \quad W_k = V_{2^k - 2}, \quad k \geq 2,$$

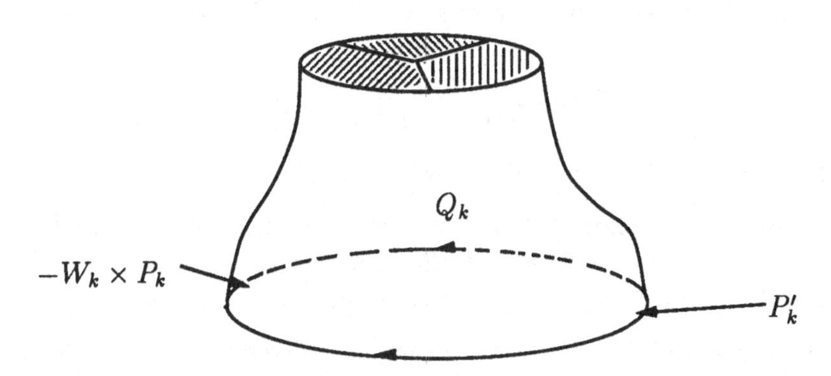

Figure 4.4: Σ_k-manifold Q_k.

we consider the Sp-manifolds

$$\widetilde{P}'_1 = R_1, \quad \widetilde{P}'_k = N_{2k-2}, \quad k \geq 2.$$

We obtain that the projection of the elements $\left[\widetilde{P}'_k\right]_{Sp}$, $k \geq 1$, into the line

$$Ext^{1,*}_{A^{MU}}(MU^*(MSp), MU^*).$$

is equal to zero due to Lemma 4.2.3. We consider the filtration of the ring MSp_*^{Σ}

$$MSp_* \xleftarrow{\gamma(1)_*} MSp_*^{\Sigma\Gamma(1)} \xleftarrow{\gamma(2)_*} MSp_*^{\Sigma\Gamma(2)} \xleftarrow{\gamma(3)_*} \cdots \tag{4.16}$$

which induces the Σ-SSS.

The elements $\left[\widetilde{P}'_k\right]_{Sp}$, $k \geq 1$, belong to the image of the homomorphism $\gamma(1)_* \circ \gamma(2)_*$:

$$\left[\widetilde{P}'_k\right]_{Sp} \in \operatorname{Im}\left(MSp_*^{\Sigma\Gamma(2)} \xrightarrow{\gamma(1)_* \circ \gamma(2)_*} MSp_*\right).$$

Any Sp-manifold Q_k, joining the manifold $\widetilde{P}'_k$ with some $\Sigma\Gamma(2)$-manifold L_k, may be considered as a Σ_k-manifold with boundary $\widetilde{P}'_k$ (see Figure 4.4).

Then

$$\partial Q_k = (\tilde{P}'_k \cup -W_k \times P_k) \cup L_k.$$

A $\Sigma\Gamma(1)$-structure on the manifold L_k is determined by the function ξ_k; see section 1.2. Dimensional reasons give $\beta_i Q_k = \emptyset$ when $i > k$. Thus we have

$$[\beta_i Q_k]_\Sigma = \left\{ \begin{array}{ll} w_k & \text{if } i = k, \\ 0 & \text{if } i \neq k, \end{array} \right\} \tag{4.17}$$

according to the definition of the manifold Q_k.

It follows from Lemma 2.2.3 that formula (4.15) holds for all $k = 1, 2, \ldots$ (here we use the fact that the numbers $p_k = \dim P_k$ are odd and that the product structure μ_k is commutative and associative).

As in Theorem 3.3.4, we suppose that for a given product structure μ the equality $w_k^2 = (4q+\epsilon)y$ holds (here $\epsilon = \pm 1$). By formula (2.8) we get

$$\beta_k^2 w_k = 2w_k + 2w_k - 4w_k = 0.$$

So $\beta_i y = 0$ for every $i = 1, 2, \ldots$. Let S be a Σ_k-manifold such that $[S]_{\Sigma_k} = q \cdot y$. We put $\tilde{Q}_k = Q_k \cup S$; then a new product structure $\tilde{\mu}_k$ determined by the product structure μ_{k-1} and the manifold $\tilde{Q}_k$ is a polynomial one (see the proof of Theorem 3.3.4). Then we have $[\beta_i \tilde{Q}_k]_\Sigma = [\beta_i Q_k]_\Sigma$ for every $i = 1, 2, \ldots$. So formula (4.15) holds. □

4.3 Some relations in the ring MSp_*

Here we begin to deal with the generators in the ring MSp_*^Σ. For this we should take some relations between the Ray elements in the symplectic cobordism ring into consideration. We are going to prove only the results that are not published yet.

Theorem 4.3.1 (V.Gorbunov [40], [41]) *The Ray elements satisfy the following relations in the ring MSp_* :*

$$\theta_1^2 \varphi_i = 0, \quad \theta_1 \varphi_i \varphi_j = 0, \quad i, j = 1, 2, \ldots. \; \square \tag{4.18}$$

Concerning the cobordism theory $MSp^*_{\theta_1}(\cdot)$ we should recall that the ring $MSp^*_{\theta_1}$ is isomorphic to the ring $\mathbf{Z}[w_1]$ down to dimension -4; here $\deg w_1 = -2$ (see Theorem 3.3.15).

Lemma 4.3.2 *The module $MSp^*_{\theta_1}(\mathbf{CP}^2)$ is free over the ring $MSp^*_{\theta_1}$.*

Proof. The second term $E_2^{*,*}$ of the Atiyah-Hirzebruch spectral sequence for $MSp^*_{\theta_1}(\mathbf{CP}^2)$ is isomorphic to the algebra

$$E_2^{*,*} \cong H^*(\mathbf{CP}^2; \mathbf{Z}) \otimes MSp^*_{\theta_1}.$$

For dimensional reasons it is clear that all the differentials in the spectral sequence are trivial. □

So we get the isomorphism

$$MSp^*_{\theta_1}(\mathbf{CP}^2) \cong H^*(\mathbf{CP}^2; \mathbf{Z}) \otimes MSp^*_{\theta_1}.$$

Let η_n be a standard linear bundle over $\mathbf{CP}^n$. We denote the bundle

$$\bar{\eta}_n \otimes \bar{\eta}_n \longrightarrow \mathbf{CP}^n$$

by Λ_n.

Now recall the *definition.* The k-dimensional real bundle ζ over X is called *symplectic orientable* if there exists a Thom class $\Delta \in MSp^k(T\zeta)$ (here $T\zeta$ is the Thom space of the bundle ζ), such that its restriction to the sphere $S^k \hookrightarrow T\zeta$ is a generator of a free MSp^*-module $MSp^*(S^k)$. In this case we have the Thom isomorphism

$$\Phi_\zeta : MSp^n(X_+) \longrightarrow MSp^{k+n}(T\zeta),$$

where $X_+ = X \cup pt$. Let $i : X \to T\zeta$ be a zero section. The element $i^*(\Delta) \in MSp^k(X)$ is called the *Euler class of the bundle* ζ.

According to [40], [41] the bundle Λ_n is symplectic orientable. In addition, its spherical fibration

$$S^1 \xrightarrow{b_n} \mathbf{RP}^{2n+1} \xrightarrow{i_n} \mathbf{CP}^n$$

is associated to the principal $U(1)$-fibration

$$S^{2n+1} \longrightarrow \mathbf{CP}^n,$$

where the action of the group $U(1)$ on the circle S^1 is determined by the homeomorphism

$$S^1 \cong U(1)/(\mathbf{Z}/2).$$

The Euler class of the bundle Λ_n is denoted by $\mathcal{P}_n \in MSp^2(\mathbf{CP}^2)$. According to [40, Theorem 5], [41, Corollary 4.8] the equalities

$$\mathcal{P}_n \cdot \theta_1 = 0, \qquad \mathcal{P}_n \cdot \varphi_j = 0, \;\; j \geq 1, \tag{4.19}$$

hold for the MSp^*-module $MSp^*(\mathbf{CP}^n)$. We denote the generator of the $MSp^*_{\theta_1}$-module $MSp^2_{\theta_1}(\mathbf{CP}^2)$ by x. As noted above the element $x^2 \in MSp^4_{\theta_1}(\mathbf{CP}^2)$ is a generator as well. These groups are connected by the following triangle:

$$\begin{array}{ccc} MSp^*(\mathbf{CP}^2) & \xleftarrow{\cdot\theta_1} & MSp^*(\mathbf{CP}^2) \\ & {\scriptstyle \pi_1} \searrow \quad \nearrow {\scriptstyle \delta_1} & \\ & MSp^*_{\theta_1}(\mathbf{CP}^2) & \end{array} \tag{4.20}$$

The proof of the following result is due to V. Gorbunov.

Theorem 4.3.3 *The equality*

$$\pi_1(\mathcal{P}_2) = 2x - w_1 x^2 \tag{4.21}$$

holds in the $MSp^*_{\theta_1}$-module $MSp^2_{\theta_1}(\mathbf{CP}^2)$.

Proof. Suppose y is a generator of the group $MSp^4(\mathbf{CP}^2)$. The Atiyah-Hirzebruch spectral sequence converging to $MSp^*(\mathbf{CP}^2)$ gives $y \cdot \theta_1 = 0$; see [43]. The triangle (4.20) gives $\delta_1 y = x$. For dimensional reasons we have $\delta_1(x^2) = 0$. The Atiyah-Hirzebruch spectral sequences for $MSp^*(\mathbf{CP}^2)$ and $MSp^*_{\theta_1}(\mathbf{CP}^2)$ give the equality

$$\pi_1(\mathcal{P}_2) = 2x + ax^2, \tag{4.22}$$

where $a \in MSp^2_{\theta_1}$. Applying the homomorphism δ_1, we have the equality $0 = 2y + \delta_1(a)y$. This holds only for the element $-w_1$, i.e. $a = -w_1$. □

Corollary 4.3.4 *The relation $\varphi_i \cdot w_1 = 0$ holds for the MSp^*-module $MSp^*_{\theta_1}$.*

Proof. The relations (4.27) imply the equality

$$\pi_1(\varphi_i) \cdot \pi_1(\mathcal{P}_2) = \varphi_i(2x + w_1 x^2) = 0.$$

Lemma 4.3.2 gives the desired relation. □

Now we come to particular generators w_k of the ring

$$MSp^{\Sigma}_* \cong \mathbf{Z}\left[w_1, \ldots, w_k, \ldots, x_2, \ldots, x_n, \ldots\right],$$

to give a geometric representation of the elements x_{2^i}.

According to Corollary 4.3.4 the Massey triple product $\langle \varphi_{2k-2}, 2, \theta_1 \rangle$ contains a zero for every $k \geq 2$. By the definition of the Massey product there exist Sp-manifolds $W_1^{(k)}$, W_k, $Y_{k,1}$, such that

$$\partial Y_{k,1} = W_1^{(k)} \times P_k \cup P_1 \times W_k.$$

We have to compare the Sp-manifolds $W_1^{(k)}$ and $W_1^{(m)}$ for different numbers k, m. The group MSp_2 contains a unique nontrivial element θ_1^2, hence the above manifolds are Sp-bordant with respect to their boundaries up to a manifold lying in the bordism class of θ_1^2.

Note 4.3.1 *If the manifolds $W_1^{(k)}$ are considered as θ_1-manifolds then they are θ_1-bordant for different k.* □

There is an important generalization of the above results; the proof below is due to V.Gorbunov.

Theorem 4.3.5 (V.Gorbunov, 1990) *For $i = 2^{q-1}$, $j > i$ each Massey product $\langle \phi_i, 2, \phi_j \rangle$ contains zero in the ring $MSp^{\Sigma_q}_*$.*

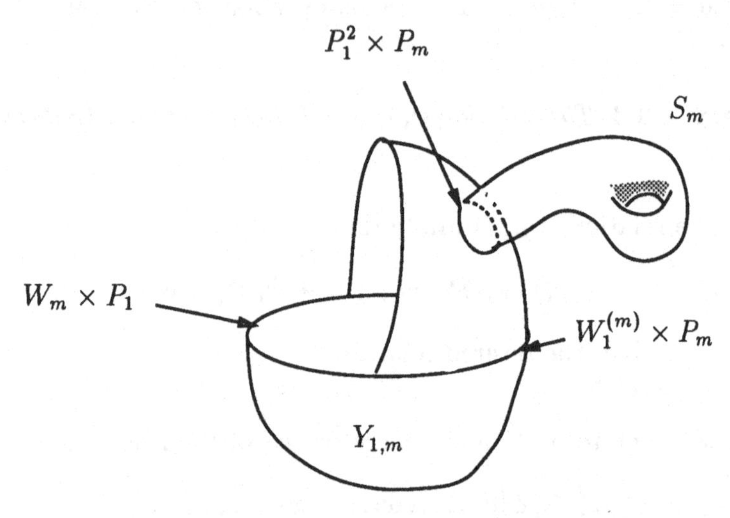

Figure 4.5: Σ_k-manifold $X_{m,1}$.

Proof. It suffices to prove that $\phi_j w_i = 0$. This was proved for $i = 1$ above. Let us assume that it's true for $j = n - 1$. We consider the diagram

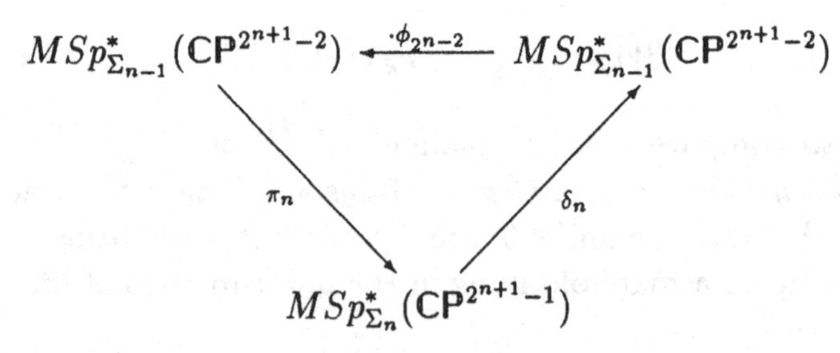

According to [41] there exists an element $d \in MSp^2(\mathbf{CP}^\infty)$ which annihilates all elements ϕ_i. We denote by d_n the element $\pi_n \cdots \pi_1(d)$. By the main result of [41] $MSp^*_{\Sigma_n}(\mathbf{CP}^{2^{n+1}-2})$ is free over $MSp^*_{\Sigma_n}$, so the element d_n may be written down as a sum $\sum_{i \geq 0} a_i z^i$, where $a_i \in MSp^*_{\Sigma_n}$. As above we conclude that $a_0 = 0$, $a_{2^{n+1}-2} = w_n$. All other coefficients a_k, $1 \leq k \leq 2^{n+1} - 2$, have the degree $4k + 2$, so they depend on w_i for $i < n$. Since the ϕ_j are of order 2, the statement of the theorem follows. □

The manifold

$$X_{m,1} = Y_{m,1} \cup -V_m \times P_m \cup S_m$$

is shown in Figure 4.5. Here the manifold S_m bounds the manifold $P_1^2 \times P_m$ (S_m exists due to Theorem 4.3.1).

The elements of the ring MSp_*^{Σ}, which are determined by the Σ-manifolds $W_1^{(2)}$, W_m, $X_{m,1}$, will be denoted by w_1, w_m, $x_{m,1}$ respectively.

4.4 Localization of bordism theories

Here we focus on the localization procedure of bordism theory with singularities. We'd like to describe it in the form which will be convenient for our purposes, putting aside its more general versions. The procedure itself will be applied in the next section.

We start with the sequence of the closed manifolds $\Sigma = (P_1, \ldots, P_k, \ldots)$ in the bordism theory $MG_*(\cdot)$. As before we suppose that the sequence Σ is locally finite, $\Sigma_k = (P_1, \ldots, P_k)$ for every $k = 1, 2, \ldots$, and $MG^{\Sigma_k}(\cdot)$ is the corresponding bordism theory with singularities. All theories $MG^{\Sigma_k}(\cdot)$ are supposed to possess admissible commutative and associative product structures compatible for different k (in the sense of Theorem 2.2.2). The pairing of the Σ_k-manifolds which generates the above product structure in the bordism theory $MG_*^{\Sigma_k}(\cdot)$ will be denoted by $\mathfrak{m}_k$. We use the notation

$$M^n = \underbrace{\mathfrak{m}_k(M, \mathfrak{m}_k(M, \mathfrak{m}_k(\cdots)\cdots))}_{n \text{ times}}$$

for all Σ_k-manifolds M. In addition, we take also the Σ_1-manifold W, $\dim W = a > 0$, which may be considered as a Σ_k-manifold for every $k \geq 2$. The element $[W]_{\Sigma_k}$ is denoted by w for every $k \geq 1$.

Finally the product homomorphism

$$MG_*^{\Sigma_k} \xrightarrow{\cdot w^n} MG_{*+an}^{\Sigma_k} \tag{4.23}$$

is supposed to be nontrivial for all numbers k and n.

Definition 4.4.1 *The Σ_k-manifold M is called a $\mathfrak{S}_{k,n}$-manifold if we have*

(i) *a partition of the boundary $\delta M = D_0 M \cup D_n M$ of the Σ_k-manifold M into the union of Σ_k-manifolds such that*

$$\delta D_0 M = \delta D_n M = D_0 M \cap D_n M$$

holds,

(ii) *a decomposition into the product of Σ_k-manifolds on the submanifold $D_n M$ of the boundary δM*

$$\Phi : D_n M \longrightarrow \mathfrak{m}_k(W^n, \mathfrak{b}^n M)$$

(i.e. Φ is a diffeomorphism preserving the G-structure). □

Definition 4.4.2 *A singular $\mathfrak{S}_{k,n}$-manifold of the pair (X, Y) is a pair (M, D_0), where M is the $\mathfrak{S}_{k,n}$-manifold, and the map*

$$f : (M, D_0 M) \longrightarrow (X, Y)$$

is a singular Σ_k-manifold such that its restriction $f|_{D_n M}$ coincides with the following composition:

$$D_n M \xrightarrow{\Phi} \mathfrak{m}_k(W^n, \mathfrak{b}^n M) \xrightarrow{pr} (\mathfrak{b}^n M)_\Sigma \xrightarrow{\widehat{f}} X.$$

Here pr is a projection on the model of the Σ_k-manifold $\mathfrak{b}^n M$, $\widehat{f}$ is a continuous map. □

Note 4.4.1 *The above notions are an obvious generalization of manifolds with singularities. According to R.Stong [103] we have that the bordism theory $MG^{\mathfrak{S}_{k,n}}(\cdot)$ of these manifolds is well defined.* □

Note 4.4.2 *The bordism theory $MG^{\mathfrak{S}_{k,n}}(\cdot)$ obviously is provided with a module structure over the theory $MG_*(\cdot)$. This module structure will mean a left module structure. In general it is not clear that the theory $MG^{\mathfrak{S}_{k,n}}(\cdot)$ has a module structure over $MG_*^{\Sigma_k}(\cdot)$. Such module structure certainly does exist if a ring structure of the spectrum MG^{Σ_k} may* *be extended up to H_∞-structure. To examine the problem seems to be interesting. Further we do not use a module structure over $MG_*^{\Sigma_k}(\cdot)$; in our case it is sufficient to know that a multiplication by w^k commutes with certain transformations.* □

Now it is convenient to consider the Bockstein-Sullivan sequence, exactness of which may be proved in the same way as in section 1.2:

$$\cdots \longrightarrow MG_*^{\Sigma_k}(\cdot) \xrightarrow{\cdot w^n} MG_*^{\Sigma_k}(\cdot) \xrightarrow{i_n} MG_*^{\mathfrak{S}_{k,n}}(\cdot) \xrightarrow{\mathfrak{b}_n} MG_*^{\Sigma_k}(\cdot) \longrightarrow \cdots \tag{4.24}$$

We obtain the following commutative diagram whose lines are exact:

$$\begin{array}{ccccccccc}
\cdots \longrightarrow & MG_*^{\Sigma_n}(\cdot) & \xrightarrow{\cdot w_1} & MG_*^{\Sigma_n}(\cdot) & \xrightarrow{\pi_1} & MG_*^{\mathfrak{S}_{1,1}}(\cdot) & \xrightarrow{\mathfrak{b}_1} & \cdots \\
& \downarrow Id & & \downarrow \cdot w_1 & & \downarrow \cdot w_1 & & \\
\cdots \longrightarrow & MG_*^{\Sigma_n}(\cdot) & \xrightarrow{\cdot w_1^2} & MG_*^{\Sigma_n}(\cdot) & \xrightarrow{\pi_2} & MG_*^{\mathfrak{S}_{1,2}}(\cdot) & \xrightarrow{\mathfrak{b}_2} & \cdots \\
& \downarrow Id & & \downarrow \cdot w_1 & & \downarrow \cdot w_1 & & \\
& \vdots & & \vdots & & \vdots & & \\
& \downarrow Id & & \downarrow \cdot w_1 & & \downarrow \cdot w_1 & & \\
\cdots \longrightarrow & MG_*^{\Sigma_n}(\cdot) & \xrightarrow{\cdot w_1^n} & MG_*^{\Sigma_n}(\cdot) & \xrightarrow{\pi_k} & MG_*^{\mathfrak{S}_{1,n}}(\cdot) & \xrightarrow{\mathfrak{b}_k} & \cdots \\
& \downarrow Id & & \downarrow \cdot w_1 & & \downarrow \cdot w_1 & & \\
\cdots \longrightarrow & MG_*^{\Sigma_n}(\cdot) & \xrightarrow{\cdot w_1^{n+1}} & MG_*^{\Sigma_n}(\cdot) & \xrightarrow{\pi_{k+1}} & MG_*^{\mathfrak{S}_{1,n+1}}(\cdot) & \xrightarrow{\mathfrak{b}_{k+1}} & \cdots \\
& \downarrow Id & & \downarrow \cdot w_1 & & \downarrow \cdot w_1 & & \\
& \vdots & & \vdots & & \vdots & &
\end{array} \tag{4.25}$$

Note that the transformation $\cdot w^n$ at the level of manifolds is the multiplication by the manifold W^n on the *left*. We have a direct limit of the diagram (4.25),

$$\cdots \longrightarrow MG_*^{\Sigma_n}(\cdot) \xrightarrow{i^0} w_1^{-1} MG_*^{\Sigma_n}(\cdot) \xrightarrow{\pi_\infty} MG_*^{\Sigma_k}/w_1^\infty(\cdot) \xrightarrow{\delta_0} \cdots$$

where

$$w_1^{-1} MG_*^{\Sigma_n}(\cdot) = \varinjlim(MG_*^{\Sigma_n}(\cdot) \xrightarrow{\cdot w_1} MG_*^{\Sigma_n}(\cdot) \xrightarrow{\cdot w_1} \cdots \xrightarrow{\cdot w_1} \cdots),$$

$$MG_*^{\Sigma_k}/w_1^{\infty}(\cdot) = \varinjlim(MG_*^{\mathfrak{S}_{1,1}}(\cdot) \xrightarrow{\cdot w_1} MG_*^{\mathfrak{S}_{1,2}}(\cdot) \xrightarrow{\cdot w_1} \cdots \xrightarrow{\cdot w_1} \cdots).$$

The theory $w_1^{-1}MG_*^{\Sigma_n}(\cdot)$ is called a *w-localization of the bordism theory* $MG_*^{\Sigma_k}(\cdot)$. The homology theories $w^{-1}MG_*^{\Sigma_k}(\cdot)$, $MG_*^{\Sigma_k}/w_1^{\infty}(\cdot)$ may be considered as ordinary bordism theories. To give a complete picture we give the definitions.

Definition 4.4.3 *An ordinary Σ_k-manifold M is called a manifold in the theory $w_1^{-1}MG_*^{\Sigma_n}(\cdot)$; the manifolds M and N are bordant in this theory if there exists a number i such that the Σ_k-manifolds $M \cdot W^i$ and $N \cdot W^i$ are Σ_k-bordant.* □

Definition 4.4.4 *A $\mathfrak{S}_{q,k}$-manifold M of dimension $m + aq$ is called a manifold of dimension m in the theory $MG_*^{\Sigma_k}/w_1^{\infty}(\cdot)$. When N is some other $\mathfrak{S}_{q',k}$-manifold of dimension $m + aq'$, then M and N are bordant in the theory $MG_*^{\Sigma_k}/w_1^{\infty}(\cdot)$, if there exists a number j such that the $\mathfrak{S}_{j,k}$-manifolds $W^{j-q} \cdot M$ and $W^{j-q} \cdot N$ are $\mathfrak{S}_{j,k}$-bordant.* □

Other attributes of the bordism theories

$$w_1^{-1}MG_*^{\Sigma_n}(\cdot) \quad \text{and} \quad MG_*^{\Sigma_k}/w_1^{\infty}(\cdot)$$

may be defined in a standard way.

Now we need some more properties of the Σ-*SSS*. The $\Sigma_k\Gamma(n)$-manifold M is glued out of the blocks $\gamma_\alpha M \times P^\alpha$ (see Definition 1.4.1), where

$$P^\alpha = P_1^{a_1} \times \ldots \times P_k^{a_k}, \quad a_1 + \ldots + a_k = n,$$

for every number collection $\alpha = (a_1, \ldots, a_k)$. A similar notion may be introduced in the case when the original bordism theory is the bordism theory with Σ_m-singularities, where $m \leq k$ (instead of the theory $MG_*(\cdot)$). In other words a Σ_m-manifold M is called a $\Sigma_{k,m}\Gamma(n)$-*manifold*, which is glued out of the blocks $\gamma_\alpha M \times P^\alpha$, where $\gamma_\alpha M$ are Σ_m-manifolds,

$$P^\alpha = P^{a_{m+1}} \times \ldots \times P^{a_k}, \quad a_{1+m} + \ldots + a_k = n,$$

for every collection $\alpha = (a_{m+1}, \ldots, a_k)$. Compatibility conditions are required here similarly to Definition 1.4.1. The complexes of the theories

$$MG_*^{\Sigma_k}(\cdot) \xrightarrow{\beta_i} MG_*^{\Sigma_k}(\cdot) \longrightarrow \cdots \longrightarrow MG_*^{\Sigma_k}(\cdot) \xrightarrow{\beta_i} \cdots \tag{4.26}$$

form the lattice of the complexes (where $i = k+1, k+2, \ldots$, signs of β_i are as in section 1.3). Suppose $\mathcal{T}^{\Sigma}_{k,m}(\cdot)$ is a total complex of this lattice:

$$MG_*^{\Sigma_{k,m}(1)}(\cdot) \xrightarrow{\beta_{k,m}(1)} MG_*^{\Sigma_{k,m}(2)}(\cdot) \xrightarrow{\beta_{k,m}(2)} MG_*^{\Sigma_{k,m}(3)}(\cdot) \xrightarrow{\beta_{k,m}(3)} \cdots$$

As before we have the transformations

$$\begin{aligned} \gamma_{k,m}(n) &: MG_*^{\Sigma_{k,m}\Gamma(n)}(\cdot) \longrightarrow MG_*^{\Sigma_{k,m}\Gamma(n-1)}(\cdot), \\ \pi_{k,m}(n) &: MG_*^{\Sigma_{k,m}\Gamma(n)}(\cdot) \longrightarrow MG_*^{\Sigma_{k,m}(n+1)}(\cdot), \\ \partial_{k,m}(n) &: MG_*^{\Sigma_{k,m}(n)}(\cdot) \longrightarrow MG_*^{\Sigma_{k,m}\Gamma(n)}(\cdot). \end{aligned}$$

Finally we come to the following theorem, the proof of which is similar to that of Theorem 1.4.2.

Theorem 4.4.5 *There is the following exact triangle of bordism theories:*

$$\begin{array}{ccc} MG_*^{\Sigma_{k,m}\Gamma(n)}(\cdot) & \xleftarrow{\gamma_{k,m}(n+1)} & MG_*^{\Sigma_{k,m}\Gamma(n+1)}(\cdot) \\ & {}_{\pi_{k,m}(n)}\searrow \quad \nearrow_{\partial_{k,m}(n+1)} & \\ & MG_*^{\Sigma_{k,m}(n)}(\cdot) & \end{array} \tag{4.27}$$

for every $n = 1, 2, \ldots$; also the equality $\pi_{k,m}(n) \circ \partial_{k,m}(n) = \beta_{k,m}(n)$ holds. □

So we have the bigraded exact couple

$$\begin{array}{ccccccc} MG_*^{\Sigma_k} & \xleftarrow{\gamma_{k,m}(1)} & MG_*^{\Sigma_{k,m}\Gamma(1)} & \xleftarrow{\gamma_{k,m}(2)} & MG_*^{\Sigma_{k,m}\Gamma(2)} & \leftarrow & \cdots \\ {}_{\pi_{k,m}(0)}\searrow & \nearrow_{\partial_{k,m}(1)} & {}_{\pi_{k,m}(1)}\searrow & \nearrow_{\partial_{k,m}(2)} & {}_{\pi_{k,m}(2)}\searrow & & \\ & MG_*^{\Sigma_{k,m}(1)} & \xrightarrow{\beta_{k,m}(1)} & MG_*^{\Sigma_{k,m}(2)} & \xrightarrow{\beta_{k,m}(2)} & \cdots & \end{array} \tag{4.28}$$

Recall that the Σ-manifold W is such that $\beta_i W = \emptyset$ for all $i = 2, 3, \ldots$. So the transformations

$$MG_*^{\Sigma_m}(\cdot) \xrightarrow{\cdot w^q} MG_{*+aq}^{\Sigma_m}(\cdot)$$

for $m \geq k+1$ may be extended up to the transformation of the exact triangles

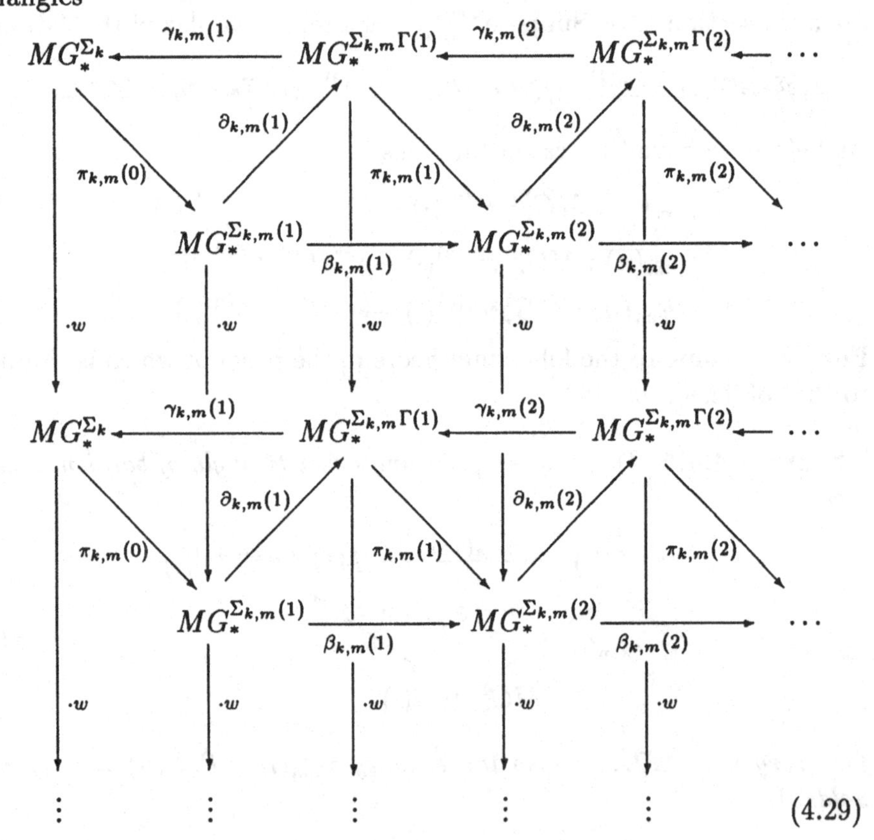

(4.29)

Setting a direct limit we obtain the exact couple

$$\begin{array}{ccccccc}
w^{-1}MG_*^{\Sigma_k} & \xleftarrow{\widetilde{\gamma}_{k,m}(1)} & w^{-1}MG_*^{\Sigma_{k,m}\Gamma(1)} & \xleftarrow{\widetilde{\gamma}_{k,m}(2)} & w^{-1}MG_*^{\Sigma_{k,m}\Gamma(2)} & \leftarrow & \cdots \\
& \searrow^{\widetilde{\pi}_{k,m}(0)} \quad \nearrow_{\widetilde{\partial}_{k,m}(1)} & & \searrow^{\widetilde{\pi}_{k,m}(1)} \quad \nearrow_{\widetilde{\partial}_{k,m}(2)} & & \searrow^{\widetilde{\pi}_{k,m}(2)} & \\
& w^{-1}MG_*^{\Sigma_{k,m}(1)} & \xrightarrow{\widetilde{\beta}_{k,m}(1)} & w^{-1}MG_*^{\Sigma_{k,m}(2)} & \xrightarrow{\widetilde{\beta}_{k,m}(2)} & \cdots &
\end{array}$$

(4.30)

The following simple lemma holds.

Lemma 4.4.6 *If the spectral sequence associated with the exact couple* (4.28) *converges to* $MG_*^{\Sigma_m}(X,Y)$ *then the spectral sequence associated with the exact couple* (4.30) *converges to* $w^{-1}MG_*^{\Sigma_m}(X,Y)$. □

In other words the operation of w-localization extends up to the operator of w-localization of the Σ-*SSS* spectral sequence and preserves its convergence.

Note 4.4.3 *The operator of introducing* $\mathfrak{S}_{n,k}$*-singularities may be extended up to the operator of the spectral sequences with the same properties.* □

Now we would like to apply the above constructions to the symplectic cobordism case.

We consider the exact couple

$$\begin{array}{ccccccc}
MSp_*^{\theta_1} & \xleftarrow{\gamma_{1,\infty}(1)} & MSp_*^{\Sigma_{1,\infty}\Gamma(1)} & \xleftarrow{\gamma_{1,\infty}(2)} & MSp_*^{\Sigma_{1,\infty}\Gamma(2)} & \leftarrow & \cdots \\
\pi_{1,\infty}(0)\searrow & \nearrow\partial_{1,\infty}(1) & \pi_{1,\infty}(1)\searrow & \nearrow\partial_{1,\infty}(2) & \pi_{1,\infty}(2)\searrow & & \\
& MSp_*^{\Sigma_{1,\infty}(1)} & \xrightarrow[\beta_{1,\infty}(1)]{} & MSp_*^{\Sigma_{1,\infty}(2)} & \xrightarrow[\beta_{1,\infty}(2)]{} & \cdots &
\end{array} \tag{4.31}$$

which induces the Σ-*SSS* for the ring $MSp_*^{\theta_1}$. In particular we have the complex

$$(MSp_*^{\Sigma_{1,\infty}(1)})_{(2)} \xrightarrow{\beta_{1,\infty}(1)} (MSp_*^{\Sigma_{1,\infty}(2)})_{(2)} \xrightarrow{\beta_{1,\infty}(2)} (MSp_*^{\Sigma_{1,\infty}(3)})_{(2)} \longrightarrow \cdots \tag{4.32}$$

which will be denoted by $\mathcal{M}$.

Now we recall some of its properties.

1. *The homology of the complex* $\mathcal{M}$ *coincides with the algebra*

$$Ext_{\mathbf{A}^{BP}}^{*,*}(BP^*(MSp^{\theta_1}), BP^*).$$

2. *The complex $\mathcal{M}$ is a total complex of the lattice the lines of which are the complexes*

$$(MSp_*^{\Sigma})_{(2)} \xrightarrow{\beta_k} (MSp_*^{\Sigma})_{(2)} \xrightarrow{\beta_k} (MSp_*^{\Sigma})_{(2)} \longrightarrow \cdots \tag{4.33}$$

for $k = 2, 3, \ldots$ (here signs of the homomorphisms β_k are as before). □

To apply the w-localization operator we put $w = w_1$, where the element w_1 is as in the section 4.3.

At first we apply the w-localization operator to the theory $MSp_*^{\Sigma}(\cdot)$. Polynomiality of the ring MSp_*^{Σ} gives the isomorphisms

$$\left(w_1^{-1} MSp^{\Sigma}\right)_* \cong w_1^{-1} MSp_*^{\Sigma}, \tag{4.34}$$

$$\left(MSp^{\Sigma}/w_1^{\infty}\right)_* \cong MSp_*^{\Sigma}/w_1^{\infty}. \tag{4.35}$$

So in this case the exact Bockstein-Sullivan sequence has the following form:

$$0 \to MSp_*^{\Sigma} \longrightarrow w_1^{-1} MSp_*^{\Sigma} \longrightarrow MSp_*^{\Sigma}/w_1^{\infty} \to 0. \tag{4.36}$$

Now let us consider the cobordism theory $w_1^{-1} MSp_*^{\theta_1}(\cdot)$. We have the Bockstein-Sullivan exact sequence

$$\cdots \to MSp_*^{\theta_1} \xrightarrow{\mathfrak{m}} (w_1^{-1} MSp^{\theta_1})_* \xrightarrow{\mathfrak{i}} (MSp^{\theta_1}/w_1^{\infty})_* \to \cdots \tag{4.37}$$

which according to the above extends to the exact sequence of complexes:

$$\begin{array}{ccccc}
\cdots \to MSp_*^{\theta_1} & \xrightarrow{\mathfrak{m}} & (w_1^{-1} MSp^{\theta_1})_* & \xrightarrow{\mathfrak{i}} & (MSp^{\theta_1}/w_1^{\infty})_* \to \cdots \\
\downarrow & & \downarrow & & \downarrow \\
0 \to MSp_*^{\Sigma_{1,\infty}(1)} & \longrightarrow & w_1^{-1} MSp_*^{\Sigma_{1,\infty}(1)} & \longrightarrow & MSp_*^{\Sigma_{1,\infty}(1)}/w_1^{\infty} \to 0 \\
\beta_{1,\infty}(1) \downarrow & & \beta_{1,\infty}(1) \downarrow & & \beta_{1,\infty}(1) \downarrow \\
0 \to MSp_*^{\Sigma_{1,\infty}(2)} & \longrightarrow & w_1^{-1} MSp_*^{\Sigma_{1,\infty}(2)} & \longrightarrow & MSp_*^{\Sigma_{1,\infty}(2)}/w_1^{\infty} \to 0 \\
\beta_{1,\infty}(2) \downarrow & & \beta_{1,\infty}(2) \downarrow & & \beta_{1,\infty}(2) \downarrow \\
\vdots & & \vdots & & \vdots
\end{array} \tag{4.38}$$

We can rewrite diagram (4.38) without the first line:

$$0 \longrightarrow \mathcal{M} \longrightarrow w_1^{-1}\mathcal{M} \longrightarrow \mathcal{M}/w_1^{\infty} \longrightarrow 0. \tag{4.39}$$

Then we have the long exact sequence

$$0 \to H_0(\mathcal{M}) \longrightarrow H_0(w_1^{-1}\mathcal{M}) \longrightarrow H_0(\mathcal{M}/w_1^{\infty}) \longrightarrow H_1(\mathcal{M})$$
$$\longrightarrow H_1(w_1^{-1}\mathcal{M}) \longrightarrow H_1(\mathcal{M}/w_1^{\infty}) \longrightarrow H_1(\mathcal{M}) \longrightarrow \cdots$$

Let $M_* = MSp_*^{\Sigma}$. Recall that we already have constructed the generators $x_{2^k-2} = x_{k,1}$ of the ring MSp_*^{Σ} for every $k \geq 2$.

Lemma 4.4.7 *The complex $w_1^{-1}\mathcal{M}$ is acyclic, i.e. $H_i(w_1^{-1}\mathcal{M}) = 0$ for every $i \geq 1$.*

Proof. Note that the complex $w_1^{-1}\mathcal{M}$ is a total complex of the lattice the lines of which are the complexes $\mathcal{L}^{(k)}$:

$$w_1^{-1}\mathcal{M} \xrightarrow{\beta_k} w_1^{-1}\mathcal{M} \longrightarrow \cdots \longrightarrow w_1^{-1}\mathcal{M} \xrightarrow{\beta_k} \cdots$$

As before it is convenient to present the complex $\mathcal{L}^{(k)}$ as a differential graded algebra $w_1^{-1}M_*[u_k]$ with a differential determined as follows:

$$\beta_k(x \cdot u_k^n) = \beta_k(x) \cdot u_k^{n+1}, \quad n = 0, 1, 2, \ldots.$$

Here u_k are generators of dimension $2(2^k - 1)$, $x \in w^{-1}M_*$. The lattice of the complexes the lines of which are the complexes $\mathcal{L}^{(k)}$, $k = 2, 3, \ldots, n$, is denoted by $\mathcal{C}(n)$, and $\mathcal{L}(n)$ is its total complex.

Lemma 4.4.8 *The complex $\mathcal{L}^{(k)}$ is acyclic for every $k = 2, 3, \ldots$.*

Proof. We have the equality

$$\beta_k\left(\frac{x_{1,k}}{w_1}\right) = 1$$

in the ring $w_1^{-1}M_*$ for every $k \geq 2$, i.e. the equality

$$\beta_k\left(\frac{x_{1,k}}{w_1}u_k^n\right) = u_k^{n+1}, \quad n = 0, 1, \ldots$$

holds for the complex $\mathcal{L}^{(k)}$. Let $x \in \mathrm{Ker}\ \beta_k$, and we obtain

$$x \cdot u_k = \beta_k \left(\frac{x_{1,k}}{w_1} x \right) \cdot u_k^n = \beta_k \left(\frac{x_{1,k}}{w_1} x \cdot u_k^{n-1} \right), \quad n \geq 1. \square$$

Lemma 4.4.9 *The complex $\mathcal{L}(n)$ is acyclic for every $n = 2, 3, \ldots$.*

Proof. The case $n = 2$ follows from Lemma 4.4.8. Let us make the induction step. The complex $\mathcal{L}(n)$ may be considered as a total complex of the double complex made up of the complexes $\mathcal{L}(n-1)$ and $\mathcal{L}^{(n)}$. The second term of the double complex spectral sequence is

$$E_2^{*,*} \cong H_I H_{II}.$$

Here H_{II} is a homology of $\mathcal{L}(n-1)$, H_I is a homology of $\mathcal{L}^{(n)}$. By induction we have $E_2^{s,*} = 0$ for every $s = 1, 2, \ldots$ and the line $E_2^{j,*}$ coincides with a homology of the complex

$$H_0\left(\mathcal{L}(n-1)\right) \xrightarrow{\beta_n} H_0\left(\mathcal{L}(n-1)\right) \xrightarrow{\beta_n} H_0\left(\mathcal{L}(n-1)\right) \longrightarrow \cdots \tag{4.40}$$

The element

$$\frac{x_{1,n}}{w_1} \in w_1^{-1} M_*$$

lies in the group $H_0(\mathcal{L}(n-1)) \subset w_1^{-1} M_*$. Actually the ring $H_0(\mathcal{L}(n-1))$ is a subring of the ring $w_1^{-1} M_*$ by induction:

$$H_0\left(\mathcal{L}(n-1)\right) = \bigcap_{k=1}^{n-1} \mathrm{Ker}\ \left(w_1^{-1} M_* \xrightarrow{\beta_k} w_1^{-1} M_* \right).$$

According to the properties of $x_{1,k}$ we have $\beta_k x_{1,n} = 0$ for every $k \neq 1, n$. So if

$$x \in H_0(\mathcal{L}(n-1)) \cap \mathrm{Ker}\ \beta_n,$$

then we get the equality

$$\beta_n \left(\frac{x_{1,n}}{w_1} x \cdot u_n^{q-1} \right) = x \cdot u_n^q, \quad q = 1, 2, \ldots,$$

so the complex (4.40) is acyclic. Thus Lemmas 4.4.9, 4.4.7 are proved. $\square$

Corollary 4.4.10 *The coefficient group of the bordism theory $w_1^{-1}MSp_*^{\theta_1}(\cdot)$ is torsion free, and there is a ring isomorphism*

$$\left(w_1^{-1}MSp^{\theta_1}(\cdot)\right)_* \cong \bigcap_{k\geq 2} \operatorname{Ker}\left(w_1^{-1}MSp_*^{\Sigma} \xrightarrow{\beta_k} w_1^{-1}MSp_*^{\Sigma}\right). \square \quad (4.41)$$

Note 4.4.4 *The bordism theory $w_1^{-1}MSp_*^{\theta_1}(\cdot)$ has an admissible product structure, and in particular the ring $(w_1^{-1}MSp^{\theta_1})_*$ is a direct limit of the rings. The computation of the ring $(w_1^{-1}MSp^{\theta_1})_*$ is to be dealt with in section 4.5.* □

Corollary 4.4.11 *There exist the exact sequence*

$$0 \to H_0(\mathcal{M}) \xrightarrow{\mathfrak{m}} H_0(w_1^{-1}\mathcal{M}) \xrightarrow{\mathfrak{i}} H_0(\mathcal{M}/w_1^{\infty}) \xrightarrow{\mathfrak{b}} H_1(\mathcal{M}) \to 0$$

and the isomorphisms

$$H_i(\mathcal{M}/w_1^{\infty}) \cong H_{i+1}(\mathcal{M}), \quad i = 1, 2, \ldots . \square$$

Note 4.4.5 *We consider the θ_1-manifold*

$$X_{1,i_1} \cdots X_{1,i_q}$$

in the bordism theory $MSp_*^{\mathfrak{S}_{1,\infty}}(\cdot)$ for every number collection $I = \{i_1, \ldots, i_q\}$. Here the manifolds X_{1,i_t} represent the elements x_{1,i_t}, and $\cdot$ denotes their θ_1-product. The equality

$$\mathfrak{b}\left(x_{1,i_1} \cdots x_{1,i_q}\right) = \sum_{t=1}^{q} u_{i_t}\left(x_{1,i_1} \cdots \widehat{x}_{1,i_t} \cdots x_{1,i_q}\right)$$

holds due to the definition of X_{1,i_t}. The projection of

$$\mathfrak{b}\left(x_{1,i_1} \cdots x_{1,i_q}\right)$$

into the term $E_2^{*,*}$ of the Σ-SSS for the bordism theory $MSp_*^{\mathfrak{S}_{\infty,1}}(\cdot)$ lies in the zero line, i.e. it maps into

$$\frac{x_{1,i_1} \cdots x_{1,i_q}}{w_1} \in MSp_*^{\Sigma}/w_1^{\infty}.$$

All higher differentials act trivially on this element since it is originally represented by a manifold in the theory $MSp_*^{\mathfrak{S}_{1,\infty}}(\cdot)$. Therefore the element

$$B(i_1,\ldots,i_q) = \sum_{t=1}^{q} u_{i_t}\left(x_{1,i_1}\cdots\widehat{x}_{1,i_t}\cdots x_{1,i_q}\right),$$

being represented by a manifold with θ_1-singularity, is a cycle of all differentials in the Σ-SSS for the ring $MSp_*^{\theta_1}$ (the same is true for the corresponding ANSS). □

Though quite obvious the above observation is very important for us.

1) The element $B(i_1,\ldots,i_q)$ coincides with the Ray element

$$\varphi_{2^{i_1}-2+\ldots+2^{i_q}-2}$$

up to decomposable elements (we mean coincidence in the ring $MSp_*^{\theta_1}$). The details will be clarified in the section 4.5.

2) The elements $B(i_1,\ldots,i_q)$ are decomposed into the Massey triple product by definition. That is, we have

$$B(i,j) \in \langle \varphi_{2^i-2}, w_1, \varphi_{2^j-2}\rangle,$$

$$B(i_1,\ldots,i_q) \in \langle B(i_1,\ldots,i_{q-1}), w_1, \varphi_{2^q-2}\rangle.$$

3) So the bordism theory $MSp_*^{\mathfrak{S}_{1,\infty}}(\cdot)$ is entirely suitable to detect torsion elements of the ring $MSp_*^{\theta_1}$. Many problems dealing with torsion may be adequately formulated in terms of this bordism theory, and interpreted as a simple version of the chromatic machinery of Ravenel; see [84]. □

4.5 The generators of the ring MSp_*^{Σ}

Recall that we have defined the *maSS*-filtration in the ring $M_* = (MSp_*^{\Sigma})_{(2)}$ which induces the *maSS*; see sections 3.3–3.5. Suppose

$$\overline{M}_* = (\overline{MSp_*^{\Sigma}})_{(2)}$$

is a ring associated with the ring M_* (with respect to the *maSS*-filtration), $\overline{\beta}_k$ is an operator associated with the Bockstein operator β_k for every $k = 1, 2, \ldots$. The polynomial generators $c_n \in \overline{M}_*$ were defined in Theorem 3.2.2, the operators $\overline{\beta}_k$ act as described in formulas (3.13)–(3.17).

Now we propose to define generators of the ring M_*, such that the action of the genuine Bockstein operators on them is similar.

We need to recall some notation.

Let $n = 2m-1$ and $m = 2^{i_1-2}+\ldots+2^{i_q-2}$ be a binary decomposition of the number m, where $2 \leq i_1 < \ldots < i_q$, $q \geq 3$; then the generator x_n is denoted by $x_{i_1,\ldots,i_q}$; if $n = 2^{i-1} + 2^{j-1} - 1$, for $1 \leq i < j$, then the generator x_n is denoted by $x_{i,j}$.

Note 4.5.1 *If the number n is even and isn't equal to a power of 2, then the generators $c_n \in \overline{M}_*$ are cycles of all differentials in the maSS. So the class c_n contains the element $x_n \in M_*$ such that the first Adams differential acts trivially on it. This means that $\beta_k x_n = 0$ for every $k = 1, 2, \ldots$.* □

The following theorem is the main result of this section.

Theorem 4.5.1 *There exist polynomial generators x_n, where $\deg x_n = 4n$, $n = 2, 4, 5, \ldots$, $n \neq 2^m - 1$, in the ring $M_* = (MSp_*^\Sigma)_{(2)}$, such that for all $k \geq 2$ the Bockstein operators β_k act on them as follows:*

1° *if $n = 2^{i-1} + 2^{j-1} - 1$, $1 \leq i < j$, then*

$$\beta_k(x_{i,j}) = \left\{ \begin{array}{ll} w_i & \text{if } k = j, \\ w_j & \text{if } k = i, \\ 0 & \text{if } k \neq i, j; \end{array} \right\} \tag{4.42}$$

2° *if $n = 2^{i_1-1} + \ldots + 2^{i_q-1} - 1$, $2 \leq i_1 < \ldots < i_q$, $q \geq 3$, then*

$$\beta_k(x_{i_1,\ldots,i_q}) = \left\{ \begin{array}{ll} w_1 \cdot x_{i_1,\ldots,\widehat{i_t},\ldots,i_q} & \text{if } k = i_t, \\ 0 & \text{if } k \neq i_1, \ldots, i_q; \end{array} \right\} \tag{4.43}$$

3° if n is an even number and is not a power of 2, then

$$\beta_k x_n = 0, \quad \textit{for all} \quad k = 2, 3, \ldots . \tag{4.44}$$

Note 4.5.2 *In particular the statement 1° of the theorem means that there exist generators $x_{1,k}$, such that $\beta_k x_{1,k} = w_1$, $\beta_i x_{1,k} = 0$ for every $i \neq k \geq 2$. Elements of this kind were determined in section 4.3. Notice that the elements $c_{1,k} = c_{2^k+1}$ are such that*

$$\overline{\beta}_1 c_{1,k} = h_k, \quad \overline{\beta}_k c_{1,k} = h_1.$$

While computing the ring $(MSp^{\Sigma}_*)_{(2)}$ (see Theorem 3.3.4) we used an arbitrary representation of the elements c_n as polynomial generators. So we can assume that the elements $x_{1,k}$ are polynomial generators of the ring M_*. □

Proof. The induction assumption is

(+) The generators x_n of the ring M_* are determined for every $n \leq N$.

Note 4.5.3 *We are going to deduce several corollaries from the assumption (+). The statements depending on the induction assumption will be marked by the sign $^{(+)}$, for example:* **Lemma 4.5.2**$^{(+)}$. □

First we examine the ring $w^{-1}M_*$ up to the corresponding dimension. That is, we consider the generators $x_{i,j}$, $x_{i_1,\ldots,i_q}$ of the ring M_* whose multi-indexes satisfy the inequalities

$$2^{i_1-1} + \ldots + 2^{i_q-1} - 1 \leq N, \quad 2^{i-1} + 2^{j-1} \leq N.$$

Lemma 4.5.2 $^{(+)}$ *The elements*

$$\begin{gathered} X_i = \frac{2x_{1,i}}{w_1} - w_i, \quad Y_i = \frac{x_{1,i}}{w_1^2}(x_{1,i} - w_i w_1), \\ X_{i,j} = x_{i,j}(x_{i,j} - w_i w_j), \quad V_i = w_i^2; \quad i = 2, 3, \ldots, \end{gathered} \tag{4.45}$$

$$X_{i_1,\dots,i_q} = x_{i_1,\dots,i_q} + \sum_{k=2}^{q-2}\left((-1)^k \sum_{1\le t_1<\dots<t_k\le q} x_{1,i_{t_1}}\cdots x_{1,i_{t_k}} x_{i_1\dots\widehat{i_{t_1}}\dots\widehat{i_{t_k}}\dots i_q}\right) \tag{4.46}$$
$$+(-1)^{q-1}\left(\frac{\sum_{t=1}^q w_{i_t}x_{1,i_1}\cdots\widehat{x}_{1,i_t}\cdots x_{1,i_q}}{w_1} - \frac{2x_{1,i_1}\cdots x_{1,i_q}}{w_1^2}\right),$$

where $q\ge 3$, $1<i_1\le\dots\le i_q$, *belong to the ring*

$$H_0(w_1^{-1}\mathcal{M}_*) \cong \operatorname{Ker}\beta_{1,\infty}(1) \subset w_1^{-1}M_*.$$

Proof. The statement concerning the elements X_i, Y_i, V_i is obvious and doesn't depend on the induction assumption (+). The statement concerning the elements $X_{i,j}$ is also obvious but it depends on (+).

As for the element $X_{i_1,\dots,i_q}\in H_0(w_1^{-1}M_*)$ the proof will be the following. We put

$$X^{(k)}_{i_1,\dots,i_q} = \sum_{1\le t_1<\dots t_k\le q} x_{1,i_{t_1}}\cdots x_{1,i_{t_k}} x_{i_1,\dots,\widehat{i_{t_1}},\dots,\widehat{i_{t_k}},\dots,i_q},$$

where $k=1,\dots,q-2$. Due to formula (4.43) we have the equality

$$\beta(1)X^{(k)}_{i_1,\dots,i_q} = w_1^{-1}\sum_{1\le t_1<\dots t_k\le q} x_{1,i_{t_1}}\cdots x_{1,i_{t_k}} A(i_1,\dots,\widehat{i_{t_1}},\dots,\widehat{i_{t_k}},\dots,i_q)$$
$$+w_1\sum_{1\le t_1<\dots<t_{k-1}\le q} x_{1,i_{t_1}}\cdots x_{1,i_{t_{k-1}}} A(i_1,\dots,\widehat{i_{t_1}},\dots,\widehat{i_{t_{k-1}}},\dots,i_q),$$

where

$$A(i_1,\dots,i_m) = \sum_{s=1}^m u_{i_s} x_{i_1,\dots,\widehat{i_{t_s}},\dots,i_m}.$$

It remains then to prove that

$$\sum_{1\le t_1<t_2\le q} x_{1,i_1}\cdots x_{1,i_{t_1}}\cdots x_{1,i_{t_2}}\cdots x_{1,i_q} A(i_{t_1},i_{t_2})$$
$$=\beta(1)\left(\frac{\sum_{t=1}^q w_{i_t}x_{1,i_1}\cdots\widehat{x}_{1,i_t}\cdots x_{1,i_q}}{w_1} - \frac{2x_{1,i_1}\cdots x_{1,i_q}}{w_1^2}\right).$$

Indeed we have due to formula (4.43)

$$\beta(1)(\cdot) = \sum_{1\leq t_1<t_2\leq q} x_{1,i_1}\cdots x_{1,i_{t_1}}\cdots x_{1,i_{t_2}}\cdots x_{1,i_q}A(i_{t_1},i_{t_2})$$

$$+2w_1^{-1}\sum_{t=1}^{q} w_{i_t}x_{1,i_1}\cdots \widehat{x}_{1,i_t}\cdots x_{1,i_q} - 2w_1^{-1}\sum_{t=1}^{q} w_{i_t}x_{1,i_1}\cdots \widehat{x}_{1,i_t}\cdots x_{1,i_q}.$$

So we have

$$\beta(1)X_{i_1,\ldots,i_q} = 0. \quad \square$$

Note 4.5.4 *We have the relation*

$$X_i^2 = 4Y_i + V_i. \quad \square \tag{4.47}$$

We consider the following subring of the ring M_*:

$$\mathcal{C}_* = \mathbf{Z}_{(2)}\left[x_n \mid n \text{ even}, \ n \neq 2^s\right].$$

So we have the isomorphism

$$w_1^{-1}M_* \cong w_1^{-1}\mathbf{Z}_{(2)}\left[w_1,\ldots,w_k,\ldots,X_i,\ldots,X_{i,j},\ldots,X_{i_1,\ldots,i_q},\ldots\right]\otimes \mathcal{C}_*,$$

which holds for the same dimensions as implied by (+). Suppose Λ_* is the polynomial ring

$$\Lambda_* = \mathbf{Z}_{(2)}\left[X_2,\ldots,X_k,\ldots,Y_2,\ldots,Y_n,\ldots\right].$$

We note that the elements V_i lie in the ring Λ_* since

$$V_i = X_i^2 - 4Y_i, \quad i = 2,3,\ldots.$$

So we can conclude:

Corollary 4.5.3 $^{(+)}$ *The isomorphism*

$$H_0(w_1^{-1}\mathcal{M}) \cong \bigcap_{k\geq 2} \mathrm{Ker}\left(w_1^{-1}M_* \xrightarrow{\beta_k} w_1^{-1}M_*\right) \tag{4.48}$$

$$\cong \mathbf{Z}_{(2)}\left[w_1,w_1^{-1}\right]\otimes \mathbf{Z}_{(2)}\left[X_{i,j},\ldots,X_{i_1,\ldots,i_q},\ldots\right]\otimes \mathcal{C}_*\otimes\Lambda_*$$

holds in the same dimensions as implied by (+). $\square$

Now we take k, such that $2^{k+1} \le N$. We can change the variables in the ring M_* as follows:

1) $x_{i,j}^{(k)} = x_{i,j}$;

2) *if* $k \notin \{i_1, \ldots, i_q\}$, *then we put*

$$x_{i_1,\ldots,i_q}^{(k)} = x_{i_1,\ldots,i_q};$$

3) *if* $k \neq i_t$, *then we put*

$$x_{i_1,\ldots,i_q}^{(k)} = x_{i_1,\ldots,i_q} - x_{1,i_t} x_{i_1,\ldots,\widehat{i_t},\ldots,i_q}.$$

Lemma 4.5.4 (+) *For* $q \ge 3$ *the generators* $x_{i_1,\ldots,i_q}^{(k)}$ *satisfy the formula*

$$\beta_m(x_{i_1,\ldots,i_q}^{(k)}) = \left\{ \begin{array}{ll} w_1 \cdot x_{i_1,\ldots,\widehat{i_t},\ldots,i_q}^{(k)} & \text{if } m = i_t \neq k, \\ 0 & \text{if } m = k, m \neq i_1, \ldots, i_q. \end{array} \right\} \qquad (4.49)$$

Proof. Let $m = i_t \neq k = i_s$. Formula (4.43) gives that

$$\beta_{i_t} x_{i_1,\ldots,i_q}^{(k)} = \beta_{i_t}(x_{i_1,\ldots,i_q} - x_{1,i_t} x_{i_1,\ldots,\widehat{i_t},\ldots,i_q})$$

$$= w_1 x_{i_1,\ldots,\widehat{i_t},\ldots,i_q} - w_1 x_{i_1,\ldots,\widehat{i_t},\ldots,\widehat{i_s},\ldots,i_q} = w_1 x_{i_1,\ldots,\widehat{i_t},\ldots,i_q}^{(k)}.$$

If $m = k = i_t$, then

$$\beta_{i_t} x_{i_1,\ldots,i_q}^{(k)} = \beta_{i_t}(x_{i_1,\ldots,i_q} - x_{1,i_t} x_{i_1,\ldots,\widehat{i_t},\ldots,i_q})$$

$$= w_1 x_{i_1,\ldots,\widehat{i_t},\ldots,i_q} - w_1 x_{i_1,\ldots,\widehat{i_t},\ldots,i_q} = 0\,. \quad \square$$

Note 4.5.5 *We note that the action of the Bockstein operator* β_i *on the generator* $x_{i,j,k}^{(k)}$ *differs from the action implied in formula* (4.49), *i.e.*

$$\beta_i x_{i,j,k}^{(k)} = w_1 x_{j,k} - w_j x_{1,k}. \quad \square$$

Lemma 4.5.5 (+) *For all numbers* $i, j, i_1, \ldots, i_q$, *satisfying* $1 \le i < j$, *and* $2 \le i_1 < \ldots < i_q$, $q \ge 3$, *there exist elements* $Y_{i,j}$, $Y_{i_1,\ldots,i_q}$ *such that*

$$\beta(1) Y_{i,j} = 2(u_i w_j + u_j w_i) = 2A(i,j), \qquad (4.50)$$

$$\beta(1) Y_{i_1,\ldots,i_q} = \sum_{s=1}^{q} u_{i_s} x_{i_1,\ldots,\widehat{i_s},\ldots,i_q} = 2A(i_1, \ldots, i_q). \qquad (4.51)$$

Proof. We put $Y_{i,j} = w_i \cdot w_j$; this case doesn't depend on (+). In the case $q = 2n > 2$ we consider the sum

$$\sum_{1 \le t_1 < \ldots < t_n \le 2n} x_{1,i_{t_1}} \cdots x_{1,i_{t_n}} x_{i_1,\ldots,\widehat{i_{t_1}},\ldots,\widehat{i_{t_n}},\ldots,i_{2n}}. \tag{4.52}$$

It is clear that for every set of numbers $\{t_1, \ldots, t_n\}$, $1 \le t_1 < \ldots < t_n \le 2n$, there exists the following complementary set:

$$\{\tau_1, \ldots, \tau_n\} = \left\{1, \ldots, \widehat{t_1}, \ldots, \widehat{t_2}, \ldots, \widehat{t_n}, \ldots, 2n\right\}.$$

So the summand

$$x_{i_{t_1},\ldots,i_{t_n}} \cdot x_{i_{\tau_1},\ldots,i_{\tau_n}}$$

enters into the sum (4.52) together with the summand

$$x_{i_{\tau_1},\ldots,i_{\tau_n}} \cdot x_{i_{t_1},\ldots,i_{t_n}}.$$

We denote

$$\sum^{S} = \frac{(-1)^n w_1}{2} \sum_{1 \le t_1 < \ldots < t_n \le 2n} x_{1,i_{t_1}} \cdots x_{1,i_{t_n}} x_{i_1,\ldots,\widehat{i_{t_1}},\ldots,\widehat{i_{t_n}},\ldots,i_{2n}}.$$

It is clear that the element

$$Y_{i_1,\ldots,i_q} = \sum_{s=1}^{2n} w_{i_s} x_{i_1,\ldots,\widehat{i_s},\ldots,i_{2n}} - \sum^{S}$$

$$-w \sum_{k=2}^{n} (-1)^k \left(\sum_{1 \le t_1 < \ldots < t_k \le 2n} x_{i_{t_1},\ldots,i_{t_k}} x_{i_1,\ldots,\widehat{i_{t_1}},\ldots,\widehat{i_{t_k}},\ldots,i_{2n}} \right)$$

satisfies formula (4.51).

Let us examine the case when q is odd. It is obvious that the element

$$Y_{i,j,k} = ((x_{i,j} w_k + x_{i,k} w_j + x_{j,k} w_i) + w_i w_j w_k)$$

satisfies formula (4.51)

Let $q = 2n+1 > 3$, $I = \{i_1, \dots, i_q\}$ and $p = i_q$. The generators $x^{(p)}_{i_1,\dots,i_q}$ are taken from Lemma 4.5.4$^{(+)}$. By their definition we get the equality

$$A(i_1,\dots,i_{2n},p) = \sum_{s=1}^{2n} u_{i_s} x_{i_1,\dots,\widehat{i_s},\dots,i_{2n},p} + u_p x_{i_1,\dots,i_{2n}}$$

$$= \sum_{s=1}^{2n} u_{i_s} x^{(p)}_{i_1,\dots,\widehat{i_s},\dots,i_{2n},p} \tag{4.53}$$

$$+ x_{1,p} \sum_{s=1}^{2n} u_{i_s} x^{(p)}_{i_1,\dots,\widehat{i_s},\dots,i_{2n}} + u_p x^{(p)}_{i_1,\dots,i_{2n}}.$$

So the element to be bounded has the following form:

$$2A(i_1,\dots,i_q) = 2\sum_{s=1}^{2n} u_{i_s} x^{(p)}_{i_1,\dots,\widehat{i_s},\dots,i_{2n},p}$$

$$+ 2x_{1,p} \sum_{s=1}^{2n} u_{i_s} x^{(p)}_{i_1,\dots,\widehat{i_s},\dots,i_{2n}} + 2u_p x^{(p)}_{i_1,\dots,i_{2n}}. \tag{4.54}$$

We put

$$Z^{(1)} = \sum_{s=1}^{2n} w_{i_s} x^{(p)}_{i_1,\dots,\widehat{i_s},\dots,i_{2n},p}, \tag{4.55}$$

$$Z^{(k)} = w_1 \sum_{1 \le t_1 < \dots < t_k \le 2n} x_{i_{t_1},\dots,i_{t_k}} x^{(p)}_{i_1,\dots,\widehat{i_{t_1}},\dots,\widehat{i_{t_k}},\dots,i_{2n},p},$$

where $k = 2, \dots, 2n-1$,

$$Z^{(0)} = Z^{(1)} + \sum_{k=2}^{2n-1} (-1)^{k+1} Z^{(k)}.$$

Formulas (4.53)–(4.55) imply

$$\beta(1)Z^{(0)} = w_1^2 \sum_{1 \le t < s \le 2n} (x_{i_t} u_{i_s} + x_{i_s} u_{i_t}) x_{i_1,\dots,\widehat{i_t},\dots,\widehat{i_s},\dots,i_{2n},p}$$

$$- w_1 x_{1,p} \sum_{1 \le t < s \le 2n} (x_{i_t} u_{i_s} + x_{i_s} u_{i_t}) x_{i_1,\dots,\widehat{i_t},\dots,\widehat{i_s},\dots,i_{2n}} \tag{4.56}$$

$$+ 2w_1 \sum_{t=1}^{2n} u_{i_t} x^{(p)}_{i_1,\dots,\widehat{i_t},\dots,i_{2n},p}.$$

Let's take

$$Z^{(*)} = x_{1,p} \sum_{s=1}^{2n} w_{i_s} x_{i_1,\dots,\widehat{i_s},\dots,i_{2n}} + w_p x_{i_1,\dots,i_{2n}},$$

$$Z^{(*,*)} = w_1 \sum_{s=1}^{2n} x_{i_s,p} x_{i_1,\dots,\widehat{i_s},\dots,i_{2n}}.$$

Then let's compute the action of $\beta(1)$:

$$\begin{aligned}\beta(1)Z^{(*)} = 2x_{1,p} \sum_{s=1} u_{i_s} x_{i_1,\dots,\widehat{i_s},\dots,i_{2n}} \\ + w_1 x_{1,p} \sum_{1\le t<s\le 2n}^{2n} (x_{i_t}u_{i_s} + x_{i_s}u_{i_t}) x_{i_1,\dots,\widehat{i_t},\dots,\widehat{i_s},\dots,i_{2n}} \\ + w_1 \left(\sum_{s=1}^{2n} w_{i_s} x_{i_1,\dots,\widehat{i_s},\dots,i_{2n}} \right) u_p + 2x_{i_1,\dots,i_{2n}} u_p + w_1 w_p \sum_{t=1}^{2n} u_{i_t} x_{i_1,\dots,\widehat{i_t},\dots,i_{2n}};\end{aligned} \tag{4.57}$$

$$\begin{aligned}\beta(1)Z^{(*,*)} = w_1 w_p \sum_{t=1}^{2n} u_{i_t} x_{i_1,\dots,\widehat{i_t},\dots,i_{2n}} + w_1 \left(\sum_{s=1}^{2n} w_{i_s} x_{i_1,\dots,\widehat{i_s},\dots,i_{2n}} \right) u_p \\ + w_1^2 \sum_{1\le t<s\le 2n} (x_{i_t}u_{i_s} + x_{i_s}u_{i_t}) x_{i_1,\dots,\widehat{i_t},\dots,\widehat{i_s},\dots,i_{2n}}.\end{aligned} \tag{4.58}$$

By comparing the equalities (4.55)–(4.58) we have

$$\beta(1) \left(Z^{(0)} - Z^{(*)} - Z^{(*,*)} \right) = 2A(i_1, \dots, i_{2n+1}). \quad \square$$

Note 4.5.6 *The elements*

$$2x_{i_1,\dots,i_q} - w_1 Y_{i_1,\dots,i_q} \in M_*$$

belong to the ring

$$\text{Л}_*^{\theta_1} = Hom^*_{\mathbb{A}^{BP}}(BP^*(MSp^{\theta_1}), BP^*),$$

in the same dimensions as implied by (+)*. These elements are adjoined to the elements* $h_0 c_{i_1,\dots,i_q}$ *in the term* $E_1^{*,*,*}$ *of the maSS for the spectrum* MSp^{θ_1}*.* $\square$

The action of the Bockstein operators β_k and the corresponding adjoint operators $\overline{\beta}_k$ are similar, as illustrated by

$$\overline{\beta}_k(h_i) = \left\{ \begin{array}{ll} h_0 & \text{if } k = i, \\ 0 & \text{if } k \neq i, \end{array} \right\} \quad \beta_k(w_i) = \left\{ \begin{array}{ll} 2 & \text{if } k = i, \\ 0 & \text{if } k \neq i, \end{array} \right\} \tag{4.59}$$

$$\overline{\beta}_k(c_{i,j}) = \left\{ \begin{array}{ll} h_i & \text{if } k = j, \\ h_j & \text{if } k = i, \\ 0 & \text{if } k \neq i, j, \end{array} \right\} \beta_k(x_{i,j}) = \left\{ \begin{array}{ll} w_i & \text{if } k = j, \\ w_j & \text{if } k = i, \\ 0 & \text{if } k \neq i, j, \end{array} \right\} \tag{4.60}$$

$$\overline{\beta}_k(c_{i_1,\ldots,i_q}) = \left\{ \begin{array}{ll} h_1 \cdot c_{i_1,\ldots,\widehat{i_t},\ldots,i_q} & \text{if } k = i_t, \\ 0 & \text{if } k \neq i_1, \ldots, i_q, \end{array} \right\} \tag{4.61}$$

$$\beta_k(x_{i_1,\ldots,i_q}) = \left\{ \begin{array}{ll} w_1 \cdot x_{i_1,\ldots,\widehat{i_t},\ldots,i_q} & \text{if } k = i_t, \\ 0 & \text{if } k \neq i_1, \ldots, i_q. \end{array} \right\} \tag{4.62}$$

The complex associated with the complex $\mathcal{M}$ with respect to the *maSS*-filtration is denoted by $\overline{\mathcal{M}}$. Comparing formulas (4.60)–(4.62) gives

Corollary 4.5.6 $^{(+)}$ *The homology algebra $H_*(\mathcal{M})$, which is isomorphic to the algebra*

$$Ext^{*,*}_{\mathbf{A}^{BP}}(BP^*(MSp^{\theta_1}), BP^*)$$

in the dimensions implied by (+), is associated to the homology algebra $H_*(\overline{\mathcal{M}})$, which is the second term of the maSS for MSp^{θ_1}. □

In other words, all higher differentials are trivial in the *maSS* for MSp^{θ_1}.

Note 4.5.7 *The above corollaries of (+) describe the structure of the ring $w_1^{-1}MSp_*^{\theta_1}$ in the corresponding dimensions, but do not provide enough information to complete the induction. We can only conclude that there exists a new generator $x_{i_1,\ldots,i_q}$ such that*

$$\beta(1)x_{i_1,\ldots,i_q} = A(i_1, \ldots, i_q) + D.$$

So we need some additional geometric information. □

We know from section 4.3 that the elements

$$B(i_1,\dots,i_q) = \sum_{t=1}^{q} u_{i_t} x_{i_1,1} \cdots \widehat{x}_{i_t,1} \cdots x_{i_q,1}$$

belong to the ring $MSp_*^{\theta_1}$ (as they are represented by θ_1-manifolds). Besides they belong to

$$H_1(\mathcal{M}) \cong Ext^{1,*}_{A^{BP}}(BP^*(MSp^{\theta_1}), BP^*)$$

by definition of the Σ-*SSS*. Let φ_m be a Ray element, where $m = 2^{i_1-1} + \dots + 2^{i_q-1}$, where $q \geq 3$. According to V.Vershinin and V.Gorbunov (see Theorem 3.3.2), the projection Φ_m of the element φ_m into the initial term of the *maSS* has the form

$$\Phi_m = \sum_{t=1}^{q} \bar{u}_{i_t} c_{i_1,1} \dots \widehat{c}_{i_t,1} \dots c_{i_q,1} + \sum_J c_J \cdot \Phi_J, \tag{4.63}$$

where $q \geq 3$, $\bar{u}_i$ is a projection of the elements θ_1, $j = 1$, and φ_{2^j-1}, $j \geq 2$, Φ_J are products of the elements Φ_n, the elements c_J belong to the ring $H_0(\mathcal{M})$ and are cycles of all differentials in the *maSS*. We recall (Theorem 3.2.2) that the element $\varphi_{i,j}$ projects to

$$\Phi_{i,j} = \bar{u}_i c_{1,j} + \bar{u}_j c_{1,i} + \sum_J c_J \Phi_J.$$

Let x_J be elements adjoined to the elements $c_J, x_J \in Л_*^{\theta_1}$. We obtain that the element

$$B(i_1,\dots,i_q) + \sum_{t=1}^{q} x_J \cdot \Phi_J \in Ext^{1,*}_{A^{BP}}(BP^*(MSp^{\theta_1}), BP^*) \tag{4.64}$$

is the image of the projection of the Ray element φ_m into the algebra

$$Ext^{1,*}_{A^{BP}}(BP^*(MSp^{\theta_1}), BP^*)$$

which is the second term E_2 of the *ANSS*. The fact that the Ray elements have order 2 and formula (4.63) allow us to conclude:

Lemma 4.5.7 *The equality*

$$2B(i_1,\ldots,i_q)=0$$

holds in the group $H_1(\mathcal{M})$. □

Now we examine the ring $Л_*^{\theta_1}$ in dimensions $4k+2$.

Lemma 4.5.8 *Every element* $x\in Л_{4*+2}^{\theta_1}$ *has the form* $x=x'w_1$, *where* $x'\in Л_{4*}^{\theta_1}$.

Proof. The cofibration

$$\Sigma^2 MSp \xrightarrow{\cdot\theta_1} MSp \xrightarrow{\pi_1} MSp^{\theta_1}$$

gives the exact sequence

$$0\to Л_* \xrightarrow{i} Л_*^{\theta_1} \xrightarrow{\beta_1} Л_* \xrightarrow{\delta} Ext^{1,*}_{A^{BP}}(BP^*(MSp),BP^*)\to\cdots \quad (4.65)$$

We remark that $Л_{4*+2}=0$. Let $x\in Л_{4*}^{\theta_1}$; then $\beta_k x=0$ for $k=2,3,\ldots$ due to the definitions. For the element $y=2x-w_1\beta_1 x$ we have

$$\beta_1 y = 2\beta_1 x - 2\beta_1 x = 0,$$

$$\beta_k y = 2\beta_k x - w_1\beta_k\beta_1 x = w_1\beta_1\beta_k x = 0, \quad k\geq 2.$$

So $y\in Л_{4*+2}=0$ and $2x=w_1\beta_1 x$. This equality holds in the polynomial ring M_*, so $x=x'w_1$. Also $w_1\beta_k x'=0$, if $k\geq 2$, i.e. $\beta_k x'=0$. So $x'\in Л_{4*}^{\theta_1}$. □

Now we are able to make a final induction step.

Suppose $x_{N+1}\in M_*$. We already have the generators for the case when $N+1$ is even.

Let $N+1=2^{i_1-1}+\ldots+2^{i_q-1}-1$. Then we take the element of the ring $w_1^{-1}M_*$,

$$Y=\frac{2x_{1,i_1}\cdots x_{1,i_q}}{w_1}-\sum_{t=1}^{q} w_{i_t}x_{i_1,1}\cdots,\widehat{x}_{i_t,1}\cdots x_{i_q,1}$$

$$+w_1\left(\sum_{k=2}^{q-2}(-1)^{k-q}\sum_{1\le t_1<\ldots<t_k\le q}x_{1,i_{t_1}}\cdots x_{1,i_{t_k}}x_{i_1,\ldots,\widehat{i_{t_1}},\ldots,\widehat{i_{t_k}},\ldots,i_q}\right).$$

Lemma 4.5.2$^{(+)}$ implies that this element has boundary

$$\beta(1)Y=(-1)^q w_1^2 A(i_1,\ldots,i_q).$$

On the other hand the homomorphism

$$w_1^{-1}MSp_*^\Sigma \longrightarrow MSp_*^\Sigma/w_1^\infty$$

maps the element Y into

$$\frac{2x_{1,i_1}\ldots x_{1,i_q}}{w_1},$$

which lies in $H_0(\mathcal{M}/w_1^\infty)$. We consider the exact sequence

$$0\to H_0(\mathcal{M})\xrightarrow{\mathfrak{m}} H_0(w_1^{-1}\mathcal{M})\xrightarrow{\mathfrak{i}} H_0(\mathcal{M}/w_1^\infty)\xrightarrow{\mathfrak{b}} H_1(\mathcal{M})\to 0\,.$$

The equality

$$\mathfrak{b}\left(\frac{2x_{1,i_1}\cdots x_{1,i_q}}{w_1}\right)=2B(i_1,\ldots,i_q)$$

holds by definition of the element $B(i_1,\ldots,i_q)$. Lemma 4.5.7 implies that there exists an element $z\in M_*$, such that

$$\beta(1)z=2B(i_1,\ldots,i_q).$$

So the element

$$z^*=Y-\frac{2x_{1,i_1}\ldots x_{1,i_q}}{w_1}-z\in M_*$$

has the boundary

$$\beta(1)z^*=(-1)^q w_1^2 A(i_1,\ldots,i_q).$$

Now we take the element $Y_{i_1,\ldots,i_q}$ from Lemma 4.3.5$^{(+)}$. The element

$$(-1)^q w_1^2 Y_{i_1,\ldots,i_q}-2z^*$$

has a zero boundary, i.e. it lies in the ring $\mathcal{J}\!\mathit{I}^{\theta_1}_{4*+2}$. Lemma 4.5.8 implies

$$(-1)^q w_1^2 Y_{i_1,\dots,i_q} - 2z^* = w_1 R,$$

where $R \in \mathcal{J}\!\mathit{I}^{\theta_1}_{4*}$. This equality holds for the polynomial ring, so the element z^* is also divided by w_1, i.e. $z^* = w_1 S$. We have

$$w_1 Y_{i_1,\dots,i_q} = 2(-1)^q S \in \mathcal{J}\!\mathit{I}^{\theta_1}_{4*},$$

so there exists the element $x_{i_1,\dots,i_q} \in M_*$, such that $\beta(1)x_{i_1,\dots,i_q} = w_1 A(i_1,\dots,i_q)$.

Now we are through with Theorem 4.5.1. □

Note 4.5.8 *Existence of the elements $x_{i,j}$, $x_{i,j,k}$ may be proved independently of the above constructions. The proof is based on the results of V.Buchstaber* [26], [27], *R.Nadiradze* [73] *and L.Ivanovskii* [46, 47]. □

So we come to the following corollaries of Theorem 4.3.1.

Corollary 4.5.9 *There are isomorphisms*

$$(\pi_* w_1^{-1} MSp^{\theta_1})_{(2)} \cong \bigcap_{k\geq 2} \mathrm{Ker}\left(w_1^{-1}M_* \xrightarrow{\beta_k} w_1^{-1}M_*\right)$$

$$\cong \mathbf{Z}_{(2)}\left[w_1, w_1^{-1}\right] \otimes \mathbf{Z}_{(2)}\left[X_{i,j},\dots,X_{i_1,\dots,i_q},\dots\right] \otimes C_* \otimes \Lambda_*,$$

where Λ_* and C_* are the following polynomial rings:

$$\Lambda_* = \mathbf{Z}_{(2)}\left[X_2,\dots,X_k,\dots,Y_2,\dots,Y_n,\dots\right],$$

$$C_* = \mathbf{Z}_{(2)}\left[x_i \mid \text{even},\ i \neq 2^s\right]. \quad \square$$

Note 4.5.9 *We note again that the generators V_i lie in the ring Λ_* :*

$$V_i = X_i^2 - 4Y_i, \quad i = 2, 3, \dots .$$

Corollary 4.5.10 *The higher differentials of the maSS for MSp^{θ_1} are trivial.* □

Note 4.5.10 *The statement of Corollary 4.5.10 was conjectured by V. Vershinin (for the spectrum MSp). The same statement is true for every MSp^{Σ_k} when $k \geq 1$.* □

4.6 Some notes

Not once has it been observed that the element $\theta_1 \in MSp_1$ possesses some specific properties in the ring MSp_*. So it seems reasonable to 'split' the problem of computing the symplectic cobordism ring. That is, we take the bordism theories with singularities $MSp_*^{\theta_1}(\cdot)$, $MSp_*^{\Sigma}(\cdot)$, $MSp_*^{\widehat{\Sigma}}(\cdot)$, where $\widehat{\Sigma} = \Sigma \setminus \{P_1\}$, to obtain the commutative square

$$\begin{array}{ccc} MSp_*(\cdot) & \xrightarrow{\pi_1^0} & MSp_*^{\theta_1}(\cdot) \\ \Big\downarrow{\scriptstyle \widehat{\pi}_1} & & \Big\downarrow{\scriptstyle \pi_1} \\ MSp_*^{\widehat{\Sigma}}(\cdot) & \xrightarrow{\widehat{\pi}} & MSp_*^{\Sigma}(\cdot) \end{array} \tag{4.66}$$

Two vertices of the square are known: these are the theories $MSp_*^{\Sigma}(\cdot)$, $MSp_*^{\widehat{\Sigma}}(\cdot)$ (the latter, being very similar to the bordism theory $MSU_*(\cdot)$, was computed in [14]). So all the main problems of the ring MSp_* are hidden in the ring $MSp_*^{\theta_1}$. The problem still remains of gluing the ring MSp_* out of the rings $MSp_*^{\theta_1}$ and $MSp_*^{\widehat{\Sigma}}$.

We have described the first Adams differential for MSp^{θ_1}. Now let's examine the result obtained more carefully. We consider the following subring of the ring $M_* = (MSp_*^{\Sigma})_{(2)}$:

$$M\langle 0\rangle_* = \mathbf{Z}_{(2)}\left[w_1, \ldots, w_k, \ldots, x_{1,2}, \ldots, x_{i,j}, \ldots, x_{i_1,\ldots,i_q}, \ldots\right].$$

We note that the ring $M\langle 0\rangle_*$ is invariant with respect to the action of the Bockstein operators β_k for every $k = 2, 3, \ldots$. We have the isomorphism

$$M_* \cong M\langle 0\rangle_* \otimes C_*,$$

where C_* is the ring defined in section 4.5. According to Theorem 4.5.1 the Bockstein operators act trivially on the ring C_*.

Now we consider the complex $\mathcal{M}$ of the previous section as a functor depending on the ring M_* with the Bockstein operators β_k, $k = 2, 3, \ldots$, acting on it. Then the complex corresponding to the ring $M\langle 0\rangle_*$ is denoted by $\mathcal{M}\langle 0\rangle$, and the trivial complex

$$\mathcal{C}_* \xrightarrow{0} \mathcal{C}_* \xrightarrow{0} \cdots \xrightarrow{0} \mathcal{C}_* \xrightarrow{0} \mathcal{C}_* \xrightarrow{0} \cdots$$

by $\mathcal{C}_*$. It is clear that

$$\mathcal{M} \cong \mathcal{M}\langle 0\rangle \otimes \mathcal{C}_*. \tag{4.67}$$

In addition we have the isomorphisms

$$Ext^{*,*}_{A^{BP}}(BP^*(MSp^{\theta_1}), BP^*) \cong H_*(\mathcal{M}) \cong H_*(\mathcal{M}\langle 0\rangle) \otimes \mathcal{C}_*.$$

Suppose

$$\mathbf{S}^{\mathbf{f}}_{\infty} = \lim_{\rightarrow q} \mathbf{S}_q$$

is the infinite symmetric group. Every group $\mathbf{S}_q$ is assumed to act on the finite set $\{2, 3, \ldots, q+1\}$, the action on other numbers being the identity. Thus the group $\mathbf{S}_q$ acts on the ring $M\langle 0\rangle_*$, the action being determined by changing indices of generators and extended over the entire ring by multiplicativity. It is clear that the direct limit $\mathbf{S}^{\mathbf{f}}_{\infty}$ of these groups acts on the ring $M\langle 0\rangle_*$ turning it into an $\mathbf{S}^{\mathbf{f}}_{\infty}$-module. We note that the product formula for Bockstein operators is invariant with respect to the action. This allows us to consider the complex $\mathcal{M}\langle 0\rangle$ as a complex of $\mathbf{S}^{\mathbf{f}}_{\infty}$-modules and to define $\mathbf{S}^{\mathbf{f}}_{\infty}$-module structure in the homology algebra $H_*(\mathcal{M}\langle 0\rangle)$. So the algebra

$$Ext^{*,*}_{\mathbf{A}^{BP}}(BP^*(MSp^{\theta_1}), BP^*) \cong H_*(\mathcal{M}\langle 0\rangle) \otimes \mathcal{C}_*$$

may be described completely in terms of the Representation Theory of the symmetric group $\mathbf{S}^{\mathbf{f}}_{\infty}$. The problem is still to be solved. One more point of interest is to figure out a geometric meaning for this $\mathbf{S}^{\mathbf{f}}_{\infty}$-module structure, i.e. to describe it in terms of characteristic classes, Two-valued Formal Groups Theory and so on.

Now let us come back to earth. We consider the first line

$$Ext^{1,*}_{\mathbf{A}^{BP}}(BP^*(MSp^{\theta_1}), BP^*) \cong H_1(\mathcal{M})$$

taken from the exact sequence

$$0 \rightarrow H_0(\mathcal{M}) \rightarrow H_0\left(w_1^{-1}\mathcal{M}\right) \rightarrow H_0(\mathcal{M}/w_1^{\infty}) \xrightarrow{\delta} H_1(\mathcal{M}) \rightarrow 0\,.$$

We determine the elements

$$Y(J; I_1, \ldots, I_n) = \frac{x_{1,j_1} \ldots x_{1,j_q} x_{I_1} \ldots x_{I_n}}{w_1}$$

of the M_*-module M_*/w_1^∞, where $J = \{i_1, \ldots, i_q\}$, and the collections I_t contain more than two elements, $I_t \cap I_s = \emptyset$, $I_t \cap J = \emptyset$ for all $s \neq t$.

It is clear that the elements $Y(J; I_1, \ldots, I_n)$ lie in the module

$$H_0 \left(\mathcal{M}/w_1^\infty \right).$$

A simple computation gives

Lemma 4.6.1 *The element*

$$A(J; I_1, \ldots, I_n) = \delta \left(Y(J; I_1, \ldots, I_n) \right)$$

has order 2^n, if $J \neq \emptyset$, and order 2^{n+1}, if $J = \emptyset$. □

It is an easy job to compute the topological dimension of the element $A(J; I_1, \ldots, I_n)$ having order 2^n. It equals $d(n) = 2^{3n+1} - (4n+7)$. For example, $d(2) = 113$, $d(3) = 1005$, $d(4) = 8169$, $d(5) = 65509$ and so on. A first such element having order 4 is

$$A_{113} = \delta \left(\frac{x_{1,5} x_{2,3,4}}{w_1} \right).$$

We consider the exact sequence

$$0 \to Л_* \xrightarrow{i} Л_*^{\theta_1} \xrightarrow{\widehat{\beta}_1} Л_* \xrightarrow{\delta} Ext^{1,*}_{ABP}(BP^*(MSp), BP^*) \to \cdots$$

generated by the Bockstein-Sullivan exact sequence

$$\cdots \to MSp_* \xrightarrow{\pi_1} MSp_*^{\theta_1} \xrightarrow{\widehat{\beta}_1} MSp_* \xrightarrow{\cdot \theta_1} MSp_* \to \cdots$$

It is clear that

$$A_{113} \in \langle x_{1,5}, w_1, x_{2,3,4} \rangle, \quad B_{111} = \widehat{\beta}_1(A_{113}) \in \langle \phi_8, 2, A_{49} \rangle,$$

where $A_{49} \in MSp_{49}$ is the element of order two, and this element of the ring $MSp_*^{\theta_1}$ has the decomposition

$$A_{49} = x_{2,3}u_4 + x_{3,4}u_2 + x_{2,4}u_3$$

at the level of the second term of the *ANSS*. In other words,

$$A_{49} = \delta\left(\frac{x_{2,3,4}}{w_1}\right).$$

Recently (September, 1989, Soviet–Japan symposium on Topology) V.Vershinin & A.Anisimov have announced the following results.

(i) *The Massey product* $\langle A_{49}, 2, \theta_1\rangle$ *contains zero.* □

(ii) *The elements*

$$B_{103} \in \langle A_{49}, 2, \varphi_7\rangle \subset MSp_{103}, \quad B_{111} \in \langle A_{49}, 2, \varphi_8\rangle \subset MSp_{111}$$

are indecomposable and have order 4. □

Result (i) means that the element $x_{2,3,4}$ can be represented by a θ_1-manifold $X_{2,3,4}$, such that

$$\delta(X_{2,3,4}) = (X_{2,3}u_4 + X_{3,4}u_2 + X_{2,4}u_3) \cdot W_1$$

where $X_{i,j}$ is a θ_1-manifold as well. To use the above geometric terms the result (ii) means that the element $A_{113} \in \langle A_{49}, w_1, \varphi_8\rangle$ has order 4 and may be represented by a θ_1-manifold. The same is true for the element $B_{111} = \delta(A_{113})$.

The element B_{103} being the first known element of order 4 has a more complicated decomposition in the *ANSS*.

The final result may be presented as follows (it is proved by elementary tools).

Theorem 4.6.2 1. *The first line*

$$Ext^{1,*}_{A^{BP}}(BP^*(MSp^{\theta_1}), BP^*)$$

of the second term of the Adams-Novikov spectral sequence has elements of order 2^n *for every* n; *the topological dimension of the first element of order* 2^n *is less than or equal to* $d(n) = 2^{3n+1} - (4n + 7)$.

2. *The algebra*

$$Ext^{*,*}_{A^{BP}}(BP^*(MSp^{\theta_1}), BP^*)$$

is multiplicatively generated by its zero and first lines. □

Note 4.6.1 September, 1991. *Recently the present author* [18] *has constructed higher torsion elements of order* 2^k *in the ring* $MSp_*^{\Sigma_2}$ *for each order* 2^k. *The first known element of order 8 in* $MSp_*^{\Sigma_2}$ *has dimension 501. The proof is based on the above geometric description of the Adams-Novikov spectral sequence for* MSp^{Σ_2} *by interpretation of V.Gorbunov's result (Theorem 4.3.5) that certain Massey products of Ray elements contain zero.*

The conclusion is that torsion elements of order 2^k *do exist in the symplectic cobordism ring* MSp_* *for all* $k \geq 2$.

Bibliography

[1] *Adams J.F.* Une relation entre groupes d'homotopie et groupes de cohomologie. C.R. Acad. Paris, 1957, T. 245, No 1, pp. 24-25.

[2] *Adams J.F.* On the structure and application of the Steenrod algebra. Comment. Math. Helv. 1958, Vol 32, pp. 180-214.

[3] *Adams J.F.* A variant of E.H. Brown's representability theorem. Topology, 1971, Vol 10, pp. 185-198.

[4] *Adams J.F.* Stable Homotopy and Generalized Homology: Univ. of Chicago Press, Chicago, Illinois, and London, 1974.

[5] *Alexander J.C.* Cobordism Massey products. Trans. Amer. Math. Soc. 1972, Vol 166, pp. 197-214.

[6] *Alexander J.C.* A family of indecomposable symplectic manifolds. Amer. J. Math. 1972, Vol 44, pp. 699-710.

[7] *Anderson D.W., Brown E.H., Peterson F.P.* SU-cobordism, KO-characteristic numbers, and the Kervaire invariant. Ann. of Math. 1966, Vol 83. pp. 54-67.

[8] *Araki S.* Typical Formal Groups in Complex Cobordism and K-theory. Lecture Notes in Math. Kyoto Univ. 6, Kinokuniya Book-Store, 1973.

[9] *Araki S., Toda H.* Multiplicative structure in mod q cohomology theories. I - Osaka J. Math. 1965. Vol 2, No 1, pp. 71 - 115, II - Osaka J. Math. 1966, Vol 3, No 1, pp. 81-120.

[10] *Baas N.A.* On the stable Adams spectral sequence. Aarhus, 1969, Preprint, Various Publ. Ser. No 6, 63 p.

[11] *Baas N.A.* On bordism theory of manifolds with singularities. Math. Scand. 1973, Vol 33, pp. 285-298.

[12] *Bakuradze M.P., Nadiradze R.G.* The cohomology realization of the two-valued formal groups and their application, Reports of the Georgian Academy , 1987, No 2, pp. 21-23 (in Russian).

[13] *Boardman J.M.* Splittings of MU and other spectra. Geometric Application of Homotopy Theory II, Proceedings. Evanston 1977. Lecture Notes in Math. No 658, pp. 28-79.

[14] *Botvinnik B.I.* Ring structure of MSU_*. Math. USSR Sbornik, 1991, Vol 69, No 2, pp. 581-596.

[15] *Botvinnik B.I.* Geometric and algebraic properties of the Adams-Novikov spectral sequence for the symplectic cobordism ring Preprint of VINITI, presented by the Siberian Math. J. No 250-1984, 59 p. (in Russian).

[16] *Botvinnik B.I.* Product structure on the cobordism with singularities. Transactions of the Math. Institute, Siberian Branch of the USSR Academy, Geometry and math. analysis, Novosibirsk, Nauka, 1986, pp. 44-61 (in Russian).

[17] *Botvinnik B.I.* The action of the Bockstein operators on the symplectic cobordism ring with the singularities. Preprint of the Computer Center, Far-East Branch of the USSR Academy, Vladivostok, 1988, 40 p. (in Russian).

[18] *Botvinnik B.I.* Higher order torsion elements in MSp_*, to appear.

[19] *Botvinnik B.I., Vershinin V.V.* On the phenomena of the initial term of the Adams-Novikov spectral sequence. Preprint of the Math. Institute, Sibirian Branch of the USSR Academy, Novosibirsk, 1981, 4 p. (in Russian).

[20] *Botvinnik B.I., Vershinin V.V.* Product structure of the singularities spectral sequence. Siberian Math. J. 1987, v.28, No 4, pp. 269-575.

[21] *Botvinnik B.I., Gorbunov V.G. Vershinin V.V.* Some applications of the spectral sequences in the Cobordism Theory. Preprint of the Math. Institute, Siberian Branch of the USSR Academy, No 26, Novosibirsk, 1986, 40 p. (in Russian).

[22] *Botvinnik B.I., Vershinin V.V.* Cobordism with singularities and spectral sequence. Colloq. Math. Soc. Janos Bolyai. Topology, Theory and Application, Budapest, 1985, Vol 41, pp. 93-117.

[23] *Brown E.H., Peterson F.P.* A spectrum whose cohomology is the algebra of reduced p^{th}-powers. Topology, 1966, V. 5, pp. 149 -154.

[24] *Browder W., Liulevicius A., Peterson F.* Cobordism theories. Ann. of Math. 1966, Vol 84, pp. 81-101.

[25] *Bruner R.* Algebraic and geometric connecting homomorphism in the Adams spectral sequence. Geometric Application of Homotopy Theory II, Proceedings, Evanston 1977, Lecture Notes in Math. No 658, pp. 131-133.

[26] *Buchstaber V.M.* Characteristic classes in cobordism and application of the two-valued and ordinary formal groups. Modern problems of mathematics, Vol 10, Moscow, 1978, pp. 5-178, (in Russian).

[27] *Buchstaber V.M.* Topological applications of the two-valued formal group theory. Math. USSR Izvestija, 1978, Vol 12, No 1, pp. 125-177.

[28] *Buchstaber V.M.* The projectors in the complex cobordism related to the SU-theory, Uspehi Math, 1972, Vol 27, No 6, pp. 231-232 (in Russian).

[29] *Buchstaber V.M.* Chern-Dold character in cobordism. Math. USSR Sbornik, 1970, No 4, pp. 573-594.

[30] *Buchstaber V.M., Novikov S.P.* Formal groups, powers series and Adams operator. Math. USSR Sbornik, 1971, Vol 13, No 1, pp. 80-114.

[31] *Cartan H., Eilenberg S.* Homological algebra, Princeton, Univ. Press, 1956.

[32] *Conner P.E., Floyd E.E.* Differential Periodic Maps. Springer, Berlin, 1964.

[33] *Conner P.E., Floyd E.E.* Torsion in SU-bordism. Mem. Amer. Math. Soc. No 60, 1966.

[34] *Conner P.E., Landweber P.S.* The bordism class of an SU-manifold. Topology, 1967, Vol 6, pp. 415-421.

[35] *Davis D.M., Mahowald M.* v_1- and v_2-periodicity in stable homotopy theory. Amer. J. Math. 1981, Vol 103, pp. 615-659.

[36] *Devinatz E.S., Hopkins M.J., Smith J.H.* Nilpotence and stable homotopy theory. Ann. of Math. 1988, Vol 128, pp. 207-242.

[37] *Dieck T.tom* Steenrod Operationen in Kobordismen-Theorie. Math. Z. 1967, Vol 107, pp. 380-401.

[38] *Douady A.* La suite spectrale d'Adams. Seminaire Henri Cartan, 11-e annee, 1958-1959, Paris, 1959, Vol 2, Exposes 18,19.

[39] *Giambolvo V., Pengelley D.C.* The homology of $MSpin$. Math. Proc. Cambr. Phil. Soc. 1984. Vol 95, p. 427-436.

[40] *Gorbunov V.G.* Symplectic cobordism of projective spaces. Math. USSR Sbornik, 1991, Vol 69, No. 2, pp. 543-557 (in Russian). also English translation.

[41] *Gorbunov V.G.* On splitting of some spectra. Preprint VINITI, presented by Sib. Math. J. No 1657 D 1986, 20 p. (in Russian).

[42] *Gorbunov V.G., Ray N.* Orientations of *Spin* bundles and symplectic cobordism. Publ. RIMS, Kyoto University; To appear 1992.

[43] *Gorbunov V.G., Vershinin V.V.* Ray's elements as the obstruction to the orientability of the symplectic cobordism theory. Soviet Math. Dokl. 1985, Vol 32, No 3, pp. 855-858.

[44] *Hazewinkel M.A.* A universal formal group and complex cobordism. Bull. Amer. Math. Soc. 1975, Vol 81, pp. 930-933.

[45] *Hopkins M.J.* Global methods in homotopy theory. Homotopy Theory. Proceedings of the Durham Symposium 1985. Cambridge Univ. Press, 1987. pp. 73-96.

[46] *Ivanovskii L.N.* Family of generators of the rational symplectic cobordism ring. Siberian. Math. J. 1986, Vol 27, No 3, pp. 358-405.

[47] *Ivanovskii L.N.* Existence of non-Stong elements in the rational symplectic cobordism. Siberian. Math. J. 1987, Vol 28, No 4, pp. 593-596.

[48] *Johnson D.C., Wilson W.S.* BP operation and Morava's extraordinary K-theories. Math. Z. 1975, Vol 144, pp. 55-75.

[49] *Kochman S.O.* The symplectic cobordism ring. I - Mem. Amer. Math. Soc. 1980, No 228; II - Mem. Amer. Math. Soc. 1982, No 271; III - Mem. Amer Math. Soc, 1992, to appear.

[50] *Kochman S.O.* The Hurewicz image of Ray's elements in MSp_*. Proc. Amer. Math. Soc. 1982, Vol 94, pp. 715-717.

[51] *Kultze R., Wurgler U.* A note on the algebra $P(n)_*(P(n))$ for the prime 2. Manuscr. Math. 1987, Vol 57, pp. 195-203.

[52] *Landweber P.S.* On symplectic bordism groups of the spaces $Sp(n)$, $\mathsf{H}P(n)$, $BSp(n)$. Mich. Math. J. 1968, Vol 15, pp. 145-153.

[53] *Landweber P.S.* Cobordism operations and Hopf algebras. Trans. Amer. Math. Soc. 1967, Vol 121, No 1, pp. 94-110.

[54] *Landweber P.S.* $BP_*(BP)$ and typical formal groups. Osaka J. Math. 1975, Vol 12, pp. 347-363.

[55] *Landweber P.S.* A survey of bordism and cobordism. Math. Proc. Camb. Phil. Soc. 1986, Vol 100, pp. 207-223.

[56] *Landweber P.S.* Homological properties of comodules over $MU_*(MU)$ and $BP_*(BP)$. Amer. J. Math. 1976, Vol 98, pp. 591-610.

[57] *Landweber P.S.* Invariant ideals in Brown-Peterson homology. Duke Math. J. 1975, Vol 42, pp. 499-505.

[58] *Lellman W., Mahowald M.* The *bo*-Adams spectral sequence. Trans. Amer. Math. Soc. 1987, Vol 300, pp. 593-623.

[59] *Liulevicius A.* Notes on homotopy of Thom spectra. Amer. J. Math. 1964, Vol 86, No 1, pp. 1-16.

[60] *Madsen Ib, Milgram R.J.* The classifying spaces for surgery and cobordism of manifolds. Princeton University Press & Univ. Tokyo Press, 1979. Annals Math. Studies No 92.

[61] *Mahowald M., Shick P.* Periodic phenomena in the classical Adams spectral sequence. Trans. Amer. Math. Soc. 1987, Vol 300, pp. 191-206.

[62] *Maunder C.R.F.* Cohomology operations of the n-th kind. Proc. Lond. Math. Soc. 1963, Vol 13, pp. 125-134.

[63] *Miller H.R.* On the relations between Adams spectral sequences, with an application to the stable homotopy of a Moore space. J. Pure and Appl. Alg. 1981, pp. 287-312.

[64] *Miller H.R., Ravenel D.C., Wilson W.S.* Periodic phenomena in the Adams-Novikov spectral sequence. Ann. of Math. 1977, Vol 106, pp. 469-516.

[65] *Milnor J.W.* On the cobordism ring Ω^* and a complex analog. Amer. J. Math. 1960, Vol 82, pp. 505-521.

[66] *Milnor J.W.* The Steenrod algebra and its dual. Ann. of Math. 1958, Vol 67, pp. 150-171.

[67] *Mironov O.K.* Existence of multiplicative structures in the theory of cobordism with singularities. Math. USSR Izvestija, 1975, Vol 9, No 5, pp. 1007-1034.

[68] *Mironov O.K.* Multiplicativity in the theory of cobordism with singularities and Steenrod-Tom Dieck operations. Math. USSR Izvestija, 1979, Vol 13, No 1, p. 89-107.

[69] *Mischenko A.S.* On the spectral sequence of the Adams type. Math. Notes of the Academy of Sci. USSR, 1967, Vol 1, No 3, pp. 226-230.

[70] *Morava J.* A product for the odd primary bordism with singularities. Topology, 1979, Vol 18, pp. 177-186.

[71] *Morava J.* Completion of complex cobordism. Lecture Notes in Math. No 658, 1978, pp. 349-361.

[72] *Morisugi K.* Massey products in MSp_* and its application. J. Math. Kyoto Univ. 1983, Vol 23, pp. 239-263.

[73] *Nadiradze R.G.* The manifolds analogous to Stong manifolds and selfconjugate cobordism. Trans. of Tbilisi Math. Inst. 1983, Vol 124, p. 65-79 (in Russian).

[74] *Nishida G.* The nilpotence of elements of the stable homotopy of spheres. J. Math. Soc. Japan, 1973, Vol 25, pp. 707-732.

[75] *Novikov S.P.* Some problems in the topology of manifolds connected with the theory of Thom spaces. Soviet Math. Dokl. 1960, Vol 29, No 1, pp. 717-719.

[76] *Novikov S.P.* On homotopy properties of Thom spaces. Mat. Sbornik, 1962, Vol 57, No 4, pp. 406-422 (in Russian).

[77] *Novikov S.P.* The methods of algebraic topology from the viewpoint of cobordism theory. Math. USSR Izvestija, 1967, Vol 1, No 4, pp. 827-913.

[78] *Nowinski K. S.* Unitary bordism with singularities determined by U_*-complex. Math. Proc. Cambr. Phil. Soc. 1984, Vol 95, pp. 443-445.

[79] *Pengelley D.C.* The homotopy type of MSU. Amer. J. Math. 1982, Vol 104, pp. 1101-1123.

[80] *Pengelley D.C.* The homology of MSO and MSU as A^*-comodule algebras and the cobordism rings. J. London Math. Soc. 1982, Vol 25, pp. 467-472.

[81] *Peterson F.P., Stein N.* Secondary cohomology operations: two formulas. Amer. J. Math. 1959, Vol 81, pp. 281-305.

[82] *Ravenel D.C.* Localization with respect to certain periodic homology theories. Amer. J. Math. 1984, Vol 106, pp. 351-414.

[83] *Ravenel D.C.* A geometric realization of the chromatic resolution. Proc. J.C.Moore Conf. Princeton. 1983. Annals Math. Studies No. 11 Princeton Univ. Press, pp. 168-179, 1987. Algebraic Topology and Algebraic K-theory.

[84] *Ravenel D.C.* Complex Cobordism and Stable Homotopy Groups. Academic Press. Orlando, Florida. 1986.

[85] *Ravenel D.C.* Formal A-modules and Adams-Novikov spectral sequence. J. Pure and Appl. Alg. 1984, Vol 32, p. 327-345.

[86] *Ray N.* Indecomposables in $Tors\,\Omega^*_{Sp}$. Topology, 1971, Vol 10, pp. 261-270.

[87] *Ray N.* Some results on generalized homology using K-theory and bordism. Proc. Cambr. Phil. Soc. 1972, Vol 71, pp. 283-300.

[88] *Ray N., Switzer R., Taylor L.* Normal structures and bordism theory with application to MSp_*. Mem. Amer. Math. Soc. No 193, 1977.

[89] *Ray N.* Bordism J-homomorphism. Ill. J. Math. 1974, Vol 18, pp. 290-309.

[90] *Ray N.* On a construction in bordism theory. Proc. Edinb. Math. Soc. 1986, Vol 26, pp. 413-422.

[91] *Roush F.W.* On the torsion classes in symplectic cobordism, Preprint. 1976.

[92] *Roush F.W.* On the symplectic cobordism ring. Comment. Math. Helv. 1970, Vol 45, pp. 159-169.

[93] *Rudjak Ju.B.* The stable K-theory and bordism with singularities. Soviet Math. Dokl. 1974, Vol 15, No 3, pp. 965-968.

[94] *Rudjak Ju.B.* Formal groups and bordism with singularities. Math. USSR Sbornik, 1975, Vol 25, No 4, pp. 487-506.

[95] *Quillen D.* Elementary proofs of some results of cobordism theory using Steenrod operations. Adv. in Math. 1971. Vol 7, pp. 29-56.

[96] *Quillen D.G.* On the formal group laws of unoriented and complex cobordism theory. Bull. Amer. Math. Soc. 1969, Vol 75, pp. 1293-1298.

[97] *Segal D.M.* On the symplectic cobordism ring. Comment. Math. Helv. 1964, Vol 45, pp. 159-169.

[98] *Shimada N., Yagita N.* Multiplication in the complex bordism theory with singularities. Publ. Res. Inst. Math. Sci. 1976-1977, Vol 12, No 1, p. 259-293.

[99] *Shimomura K., Yoshimura Z.* BP-Hopf module spectrum and BP_*-spectral sequence. Publ. RIMS, Kyoto Univ. 1986, Vol 21, pp. 925-947.

[100] *Smith L., Zahler R.* Detecting stable homotopy classes by primary BP operations. Math. Z. 1972, Vol 129, pp. 137-156.

[101] *Snaith V.P.* Algebraic cobordism and K-theory. Mem. of Amer. Math. Soc. No 221, 1979.

[102] *Steenrod N.E., Epstein D.B.A.* Cohomology operations. Ann. Math. Studies No 50, 1962, Princeton University Press.

[103] *Stong R.* Notes on cobordism theory. Princeton University Press, Princeton, New Jersey, 1968.

[104] *Stong R.E.* Some remarks on symplectic cobordism. Ann. of Math. 1967, Vol 86, pp. 425-433.

[105] *Sullivan D.* The Hauptvermutung for manifolds. Bull. Amer. Math. Soc. 1967, Vol 73, pp. 598-560.

[106] *Sullivan D.* Geometric periodicity and the invariants of manifolds. Manifolds. Proceedings. Amsterdam. 1970, Berlin-Tokyo, 1971, pp. 44-75.

[107] *Sullivan D.* Geometric topology. Localization, periodicity and Galois symmetry. MIT, Cambridge, Massachusetts, 1970.

[108] *Switzer R.M.* Algebraic topology - homotopy and homology. Springer-Verlag, 1975.

[109] *Vershinin V.V.* Algebraic Novikov's spectral sequence for the spectrum MSp. Siberian Math. J. 1980, Vol 21, No 1, pp. 19-31.

[110] *Vershinin V.V.* Computation of the symplectic cobordism ring below the dimension 32 and nontriviality of the majority of triple products of the Ray elements. Siberian Math. J. 1983, Vol 24, No 1, pp. 41-51.

[111] *Vershinin V.V.* Symplectic cobordism with singularities. Math. USSR Izvestija, 1984, Vol 22, No 2, pp. 211-220.

[112] *Vershinin V.V.* Cobordism and spectral sequence. Doctoral dissertation, Novosibirsk, 1988, 210 p. (in Russian). English translation to be published by AMS.

[113] *Vershinin V.V.* Multiplication in the spectral sequences. Transactions of the Math. Institute, Siberian Branch of the USSR Academy, Geometry and math. analysis, Novosibirsk, Nauka, 1986, pp. 61-70 (in Russian).

[114] *Wall C.T.C.* Addendum to a paper of Conner-Floyd. Proc. Cambr. Ph-l. Soc. 1966, Vol 62, pp. 54-67.

[115] *Wilson W.S.* The Hopf ring for Morava K-theories. Publ. RIMS, Kyoto Univ. 1984, Vol 20, pp. 1025-1036.

[116] *Wurgler U.* On the products in a family of cohomology theories associated to the invariant prime ideals of $\pi_*(BP)$. Comment. Math. Helv. 1977, Vol 52, pp. 457-481.

[117] *Wurgler U.* A splitting for certain cohomology theories associated to $BP^*(\cdot)$. Manuscr. Math. 1979, Vol 29, pp. 93-111.

[118] *Wurgler U.* Cobordism theories of unitary manifolds with singularities and formal group laws. Math. Z. 1976, Vol 150, pp. 239-260.

[119] *Wurgler U.* Commutative ring-spectra of characteristic 2. Comment. Math. Helv. 1986, Vol 61, pp. 33-45.

[120] *Yagita N.* On the algebraic structure of cobordism operations with singularities. J. London Math. Soc. 1977, Vol 16, pp. 131-141.

[121] *Yagita N.* On the Steenrod algebra of Morava K-theory. J. London Math. Soc. 1980, Vol 22, pp. 423-438.

[122] *Yagita N.* A topological note on the Adams spectral sequence based on Morava's K-theory. Proc. Amer. Math. Soc. 1978, Vol 72, pp. 613-617.

[123] *Zahler R.* The Adams-Novikov spectral sequence for spheres. Ann. of Math. 1972, Vol 96, pp. 480-504.

[124] *Zahler R.* Detecting stable homotopy with secondary cobordism operations. Quart. J. Math. Oxford 1974, Vol 25, pp. 213-226.

Printed in the United States
By Bookmasters